T0076338

PRAISE FOR *WHO's C*
UNITING NUMBERS AND NARRATIVES WITH STORIES FROM
POP CULTURE, PUZZLES, POLITICS, AND MORE

"*Who's Counting?* is an astonishing book, for it combines two things that aren't often found together: analytical rigor and fun. You don't have to be a math nerd to appreciate the fluid, easy explanations of everyday innumerate reasoning, but I guarantee that this book will make you feel smarter. *Who's Counting?* is structured to enable the reader to dip in and out, biting each chocolate to see what one likes, but I encourage you to do what I did and just gorge yourself. It all tastes good."—**Lee McIntyre, author of *How to Talk to a Science Denier***

"Math illiteracy breeds warped understandings of the world that ultimately jeopardize the progress of civilization. Math professor John Allen Paulos wrote *Who's Counting?* as an entertaining, relevant, and potent antidote to this societal blight."—**Neil deGrasse Tyson, astrophysicist, American Museum of Natural History**

"John Allen Paulos yet again defends logic and crusades against the problem of innumeracy with good humor and plain language. Using examples ranging from math puzzles to controversial cultural issues, this engaging book reminds us that citizens in a democracy must understand the facts and figures that swirl about them every day."—**Tom Nichols, author of *The Death of Expertise***

"What better than a mathematician sharing the intrigue of math with us all and, at the same time, protecting us from the data manipulation and fake news that is threatening our democracy. *Who's Counting?* is a must-read for all who want to wear mathematical armor against the war on truth."—**Jo Boaler, Nominelli-Olivier Professor of Education (Mathematics) at Stanford University, co-founder of youcubed.org, and author of *Limitless Mind***

WHO'S COUNTING?

Uniting Numbers and Narratives with Stories from
Pop Culture, Puzzles, Politics, and More

JOHN ALLEN PAULOS

Prometheus Books
Essex, Connecticut

 Prometheus Books

An imprint of Globe Pequot, the trade division of
The Rowman & Littlefield Publishing Group, Inc.
4501 Forbes Blvd., Ste. 200
Lanham, MD 20706
www.rowman.com

Distributed by NATIONAL BOOK NETWORK

British Library Cataloguing in Publication Information Available

Library of Congress Cataloging-in-Publication Data

Names: Paulos, John Allen, author.
Title: Who's counting? : uniting numbers and narratives with stories from pop culture, sports,
    politics, and more / John Allen Paulos.
Description: Lanham, MD : Prometheus, an imprint of Globe Pequot, the trade division of The
    Rowman & Littlefield Publishing Group, Inc., [2022] | Summary: "Who's Counting features
    selected columns from Paulos's well-known ABC News series of the same name collected
    here in book form for the first time, along with updates and brand-new original essays from
    the author, to examine how better understanding data improves our thinking and decision-
    making"—Provided by publisher.
Identifiers: LCCN 2022001235 (print) | LCCN 2022001236 (ebook) | ISBN 9781633888128
    (paperback) | ISBN 9781633888135 (epub)
Subjects: LCSH: Mathematics—Popular works.
Classification: LCC QA93 .P374 2022 (print) | LCC QA93 (ebook) | DDC 510—dc23/
    eng20220521
LC record available at https://lccn.loc.gov/2022001235
LC ebook record available at https://lccn.loc.gov/2022001236

*Dedicated to those who can be both
tough-minded and softhearted*

Contents

INTRODUCTION

MORE THAN 30 YEARS AGO, I WROTE A WELL-RECEIVED BOOK TITLED *Innumeracy*. It contained some basic mathematical ideas, a few pedagogical suggestions, a bit of irreverence, and a cataloging of the consequences of Americans' quantitative ignorance, at least those of a considerable fraction of us. Since then, in part a result of the wide availability of countless apps, software, books, and media platforms, people have become a little more comfortable with numbers and statistics. And perhaps because of the very welcome focus on so-called STEM subjects (science, technology, engineering, and mathematics) in schools and in the general culture, more people are also familiar with at least some mathematics and hopefully have grown accustomed to the ubiquitous uses of the subject.

Yet, despite all this, one of *Innumeracy*'s primary leitmotifs still holds today: Too many of us are still innumerate, and this societal innumeracy remains a vastly underrated driver of bad policy and bad politics. Almost every major issue and many minor ones facing this country are made more difficult by believing that (or at least acting as if) the numbers, probabilities, and relative magnitudes relevant to them don't really matter. Unfortunately, the problem involves not only the formal properties of these figures but also an understanding of what they mean or, more frequently, don't mean and how they should be interpreted.

For example, if you add 2 cups of water to 2 cups of popcorn, you don't get 4 cups of soggy popcorn, even though 2 plus 2 is unassailably equal to 4. Perhaps an absurd example, but it does point to a kind of extra-mathematical contextual understanding that is necessary but very often lacking in the application of mathematics to many topical issues. Why is it, to cite a less artificial recent example, that in highly vaccinated states, almost half of those hospitalized have been fully vaccinated? This seemingly odd fact is no more difficult to explain than is the following: All the students in a certain region took a difficult exam. Almost half of those who failed the exam studied hard for it. This doesn't mean studying hard for it was useless, merely that a huge majority of those taking the exam studied hard for it.

When we use mathematics to describe the social world, we're always faced with a number of questions. What are we trying to measure or count, and how do we decide whom to include, exclude, or qualify? For example, is an unemployed person who takes turns staying with different siblings really homeless? And what degree of importance, certainty, or relevance should we ascribe to relationships between quantities, say, the relationship between COVID-19 and poverty or between Facebook and conspiracy theories? In addition to being fun, as I hope the examples included herein illustrate, precisely stated mathematical definitions and conditions can sometimes help reveal the implicit assumptions underlying such questions.

There are, of course, many illustrations of these complicated issues over the past few years, some deriving from Trump and COVID-19. Despite the latter two mega- (or maga-) topics, the columns I have selected from the archive of my monthly ABCNews.com Who's Counting columns, having been written between 2000 and 2010, necessarily focus on other issues. They discuss a multiplicity of topics, in part because mathematics is a somewhat imperialist discipline that regularly invades other disciplines.

Also, since all the mathematical, and many of the social issues, discussed in the columns are relatively independent of time, they're still informative despite superficial anachronisms that I beg the reader to ignore. At times, they even provide a smidgeon of historical context.

The book may be considered a mathematical sampler. (I hope it's not considered a box of stale bon bons, but I leave that to the reader.) Some of the columns will be relevant to public policy questions and will involve misconceptions that are unfortunately perennial. Another group, perhaps more fun, will deal with puzzles and paradoxes since an appreciation of recreational mathematics and a feel for its sometimes counterintuitive conclusions is likely to make one more sensitive to the beauty and even the utility of mathematics. Still another group will offer a conception of mathematics broader than that of a discipline devoted solely to calculation and theorem-proving.

I also include commentaries on the columns, some quite extended, as well as a good number of more recently written pieces that are not Who's Counting columns and that deal primarily with contemporary matters. Many illustrate Samuel Johnson's quip, "A thousand stories which the ignorant tell, and believe, die away at once when the computist takes them in his gripe." The spelling is a bit antiquated, but the sentiment is just as apt as it ever was.

Obviously, no one expects students, citizens, legislators, or government leaders to understand the arcsine law or the Banach–Tarski theorem. They should, however, know the difference between risk and relative risk for various subgroups of the population, for example, or the difference between the mean of a set of numbers and its median, say, between the mean income and the median income in the United States, or the basics of scaling physical and other quantities up or down.

Clarifying, at least a little, the mathematical aspects of these various concerns is a bit like taking out the garbage. You can't just do it once since the garbage tends to pile up and proliferate. It has to be an ongoing practice, which I immodestly aver is a large part of the rationale for this book and its inclusion of older but still germane writings of mine. In addition, I've not been able to resist briefly revisiting old peeves and obsessions. Nonsensical mathematical claims can be repeated over and over, but there's an often unfortunate reluctance to repeat sensible explications of mathematical notions and their applications such as those, I like to think, that appear in my writings. I certainly can't discuss even a minuscule fraction of social and popular issues with a significant mathematical flavor, so I've made my own idiosyncratic choices. Happily, there are a number of very good garbage-collecting writers out there as well.

The broad topics discussed are puzzles, a bit of probability, lies and logic, (mis)calculations, political partisanship, and religious dogmatism. Among the specific topics discussed are the following: What does Lanchester's law have to do with guerrilla warfare? How can two losing strategies be combined into a winning strategy? What is the relation between "sexonomics" and prostitution? How can Kruskal's card paradox be used to dupe the gullible and explain "happiness"? Did a 13th-century German monk discover the Mandelbrot set long before Benoit Mandelbrot did? Is denying a bit of disinformation always a good strategy? What does Wolf's dilemma reveal about political parties' unity? What is complexity, and how can it illuminate an intuition behind Godel's famous incompleteness theorem? Should we require that presidential candidates solve some very basic math problems, and of what sort? How is it that the number e, 2.718281 . . . , pops up in so many common situations? What is the relationship between the quantitative and the narrative aspects of a story? Should Joe DiMaggio's hitting streak have an asterisk attached? How can the right metaphors increase our understanding of mathematical notions? What is a fundamental flaw in probability that vitiates creationist arguments? What do coincidences, apophenia, and coin flips say about eyewitness testimony, blood clots, and any number of other issues? And, as sales pitches often annoyingly promise, much more.

In passing, there will be asides about many matters, one of the most simple being large numbers. How much credibility, for example, would you accord a journalist, politician, or citizen who regularly confused events that occurred one or two weeks ago with events that occurred in 1988 or sometimes even with events that occurred in the Upper Paleolithic era? Sounds absurd, but as I've often repeated, this is the difference between 1 million seconds (11.5 days), 1 billion seconds (32 years), and 1 trillion seconds (32,000 years), and confusions like this are still common. These are huge numbers, of course, but almost nothing compared to the number of ways in which the 52 cards in a deck could be arranged in order, which is a bit more than 8×10^{67}, 8 followed by 67 zeros. By comparison, there are only about 7×10^{14} seconds in the approximately 14-billion-year history of the universe. No one has a visceral feel for numbers this gargantuan.

The approach throughout will be based on stories and vignettes and puzzles that help explain, contextualize, and vivify the relevant mathematical notions. There won't be formulas or equations, although occasionally they will be implicit in the stories. Some of the complexities of conditional probability, for example, are illustrated in the following story and its explication in one of the columns. If at least one of a woman's two children is a boy, the probability she has two boys is 1/3, but if at least one of a woman's two children is a boy born in summer, the probability she has two boys is 7/15. Why? It can be easily explained, and we need no elaborate theory involving global warming or genetics or anything else to account for the above probability.

This brings me to elaborate theories, conspiracy theories, which often depend on an inability to explain or evaluate the likelihood of events. The internet makes so many facts of dubious origin available to us that confirmation bias, partisanship, and the so-called conjunction fallacy, the tendency we have to attribute plausibility to scenarios with many extraneous details, all help account for the rise in "fake news," such as the claimed diabolical intent behind vaccines.

Finally, it is important that we (and, of course, I) not adopt too earnest and scolding an attitude toward innumerate beliefs and attitudes. I am, after all, an aging baby boomer who is beginning to lose control of his eyebrows and thus more prone to such an attitude. Everybody is wrong sometimes, and our desire for certainty is almost always unsatisfiable, but as I wrote in *Innumeracy*, numbers, probabilities, and logic are, along with a humble respect for truth, our most basic and reliable guides to reality. A society that replaces them with power, wealth, and duplicity is not a healthy one.

CHAPTER ONE

Puzzles as a Prelude

Mathematical problems, or puzzles, are important to real mathematics (like solving real-life problems), just as fables, stories, and anecdotes are important to the young in understanding real life.

—TERENCE TAO

AS I WROTE IN MY FIRST BOOK, *MATHEMATICS AND HUMOR*, BOTH MATHematics and humor are activities undertaken for their own sake. In both, ingenuity and cleverness are prized. Of course, I am speaking here of pure mathematics, the art and science of abstract pattern and structure, and not of computational mathematics, which is to a certain extent a collection of very useful techniques and algorithms. I am also referring to "pure humor." The analogue to computational mathematics might be, I suppose, manipulative uses of humor in public relations, advertising, and promotion.

Logic, patterns, rules, and structure are also essential to both mathematics and humor albeit with a different emphasis. In humor the logic is often inverted, patterns are distorted, rules are misunderstood, and structures are confused. Yet these transformations are not random and must still make sense on some level. Understanding the "correct" logic, pattern, rule, or structure is essential to understanding what is incongruous in a given story, to "getting the joke."

In addition, both mathematics and humor are economical and explicit. Thus, the beauty of a mathematical proof depends to a certain extent on its elegance and brevity. A clumsy proof introduces extraneous considerations; it is long winded or circuitous. Similarly, a joke loses its humor if it is awkwardly told, is explained in redundant detail, or depends on strained analogies.

The logical technique of reductio ad absurdum also often plays a role in both humor and mathematics. It is a favorite gambit in mathematical proofs and, simply stated, comes to the following. To prove statement S, it is enough to assume the negation of S (not S) and from the negation derive a contradiction. And jokes regularly depend on the literal interpretation of statements to stress a humorous incongruity.

But somewhere in the vague middle of the pure math/pure humor continuum lie puzzles. They're often more substantial than mere jokes yet can also reveal a significant mathematical fact. The analogue of a joke's punch line is the "aha" moment when a surprising insight dawns on us, and the thought process required to solve a puzzle often mirrors that in a mathematical analysis. Not exactly math and not exactly humor, puzzles often provide a gateway to real mathematics as the works of puzzle-meister Martin Gardner and others amply demonstrate. In fact, Gardner was probably at least partly responsible for gently luring many mathematicians into the discipline. The title of one of his books encapsulates much of his work, *Aha Insight*.

Sudden insight brings to mind a comment by the philosopher Ludwig Wittgenstein, who once wrote that a good and serious work in philosophy could be written that consisted entirely of jokes (interpreted quite broadly to include puzzles and anecdotes). The idea is that if one gets the joke, one gets the relevant philosophical point. My book *I Think, Therefore I Laugh*, attempted to illustrate Wittgenstein's contention about philosophical ideas. Furthermore, in the same way that jokes can provide insights into philosophical problems and conundrums, I believe that mathematical puzzles can be an effective means to explore and vivify mathematical notions.

For these partially personal reasons, as well as the fact that puzzles are fun, I've included herein a number I've discussed in my ABCNews. com columns throughout the years. Some appear in this first section, but less overtly mathematical examples appear throughout the book.

The counterintuitive Monty Hall problem continues to baffle people if the emails I still receive insisting that the sender's proposed solution is correct is any indication. I sometimes feel like the advice columnist in *Miss Lonelyhearts* as I try to gently point out their errors. Herein I try to make the problem's solution plausible and introduce variants of it that might be helpful.

A metapsychological problem is to understand why so many people are unconvinced by all of the various solutions. Sometimes people even cite the large number of the unconvinced as proof that the solution is a matter of real controversy, just as in politics an inconvenient fact such as global warming is obscured by fake controversies. Having mentioned an inconvenient (to some) fact, I note below a surprising connection between the Monty Hall problem and a COVID-19 precaution.

THE MONTY HALL PUZZLE: VARIANTS OF IT AND ITS CONNECTION TO A COVID-19 PRECAUTION

Many new game shows have appeared in recent years, among them *Who Wants to Be a Millionaire*, *Deal or No Deal*, and *Show Me the Money*. So far, none has aroused the mathematical interest of the quiz show *Let's Make a Deal*. Having received so many emails throughout the years about the show's so-called Monty Hall problem, I thought I'd devote this holiday column to the famous problem, to a question about a new variant of it (with answer at the end), and to a more general question about its applicability.

Let's Make a Deal was popular in the 1960s and 1970s and has been resurrected in various formats since then. In its original version the game show contestant was presented with three doors, behind one of which is a new car. The other two doors have booby prizes behind them.

The Original Problem

The *Let's Make a Deal* host, Monty Hall, asks the contestant to pick one of the three doors. Once the contestant has done so, Monty opens one of the two remaining doors to reveal what's behind it but is careful never to open the door hiding the promised new car. After Monty has opened one of these other two doors, he offers the contestant the chance

3

to switch doors. The question is, Should the contestant stay with his (or her) original choice and hope the car is behind it or switch to the remaining unopened door?

Many people reason that it doesn't make any difference since there are two possibilities, and thus the probability is 1/2 that the car is behind the original door. The correct strategy, however, calls for the contestant to switch. The probability that he picked the correct door originally is 1/3, and the probability that the car is behind one of the other two unopened doors is 2/3. Since the host is required to open a door behind which there's a booby prize, the 2/3 probability is now concentrated on the other unopened door. Switching to it will increase the contestant's chances of winning from 1/3 to 2/3.

One common way to make the decision to switch more intuitively appealing is to imagine that there are 100 doors, behind one of which is a new car and behind the other 99 of which are booby prizes. The host again asks the contestant to pick one of the doors. Once the contestant has done so, the host opens 98 of the 99 remaining doors to reveal what's behind them but is careful never to open the door hiding the car. After the host has opened 98 of these other 99 doors, he offers the contestant the chance to switch to the other unopened door. The question again is, Should the contestant switch?

That switching is the correct answer is clearer in this case. The probability that the contestant picked the correct door originally is 1/100, and the probability that the car is behind one of the other 99 unopened doors is 99/100. Since the host is constrained to open 98 doors behind which there's a booby prize, the 99/100 probability is now concentrated on the other unopened door. Switching to it will increase the contestant's chances of winning from 1/100 to 99/100.

It's interesting that even in the latter case many people refuse to switch. One factor may be an extreme fear of the regret they'd feel if they switched away from their original pick and it happened to be correct. The same fear of regret underlies people's reluctance to trade lottery tickets with friends. They imagine how bad they'd feel if their original ticket were to win.

A New Variant, a General Question, and the
Connection between Monty Hall and a COVID-19 Precaution

If this is all clear, here's a variation of the problem to test your understanding of the probability involved. (The answer is below.) Let's say that *Let's Make a Deal* were to attempt a comeback. The producers, fearing the audience would be small if the game were exactly the same, devise a variant game in which the contestant is presented with 10 doors. Again, behind one of them is a car, and behind the others are booby prizes. After the contestant picks a door, Monty (or his avatar) opens just seven of the remaining nine unopened doors but is careful never to open the door hiding the car. There are now three unopened doors: the one that the contestant originally picked and two others. Which strategy works best: switching to one of the other two unopened doors or sticking with the original pick? Furthermore, what is the probability of winning by following these two strategies?

Answer: The chance the prize is behind the door originally chosen by the contestant is 1/10 and remains 1/10. The chance it's behind one of the nine other unopened doors is 9/10. Since the host opens seven of these nine other unopened doors, the 9/10 probability that it's behind one of them is concentrated on and divided between two of these nine unopened doors. So the contestant should switch to one of these two. Doing so raises his probability of winning from 1/10 to one-half of 9/10, or 45%.

One last question: Can you think of any situations—crime mysteries, world politics, imagined scenarios—that might be modeled on some close variant of the Monty Hall problem?

Answer: Not surprisingly, my response is also the reason I posed the question. As promised, I maintain that the Monty Hall problem has a natural connection to COVID-19. Consider what mathematicians call a dual problem, which is a somewhat different but essentially identical problem whose solution provides or clarifies the solution of the original.

Let's assume that the "prize" is not a new car but an extraordinarily infectious virus that the contestant wants to avoid. He is told that the virus is spring-loaded behind exactly one of the doors and is forced by

Taunty Hall, Monty's sadistic cousin, to pick one of the three doors. Once the contestant has done so, the substitute host Taunty, who has reluctantly agreed never to open the door hiding the virulent virus, opens one of the two unpicked doors to reveal what, if anything, is behind it. After Taunty has opened one of the other two doors, he offers the contestant the chance to switch his or her choice. The question is, Should the contestant stay with the original choice of door and hope nothing is behind it or switch to the remaining unopened door?

The same analysis as that provided for the original problem holds true here, but this time it counsels the contestant to stay with his original choice rather than switch. If he does stay, his probability of being exposed to the virus is and remains 1/3, and so the probability that the virus was placed behind one of the other two unopened doors is 2/3. Since the host has agreed to open a door behind which there is nothing, the 2/3 probability is now concentrated on the other unopened door. Switching to it will thus increase the contestant's chances of choosing the door with the spring-loaded virus from 1/3 to 2/3.

The COVID-19 version is, as noted, the dual problem to the original, but I think the correct response to it is more intuitive than the correct response to the original version. It also underscores the wise preference to limit one's exposure to the virus to as few people as possible or, in the case above, to one person rather than two.

More generally, I would argue that puzzles often provide a logical skeleton of real-life situations.

It's well known that businesses test potential employees with simple logic puzzles. The natural question arises: Why not put presidential candidates to the test? The questions shouldn't be arcane or require special knowledge but rather ones that require clear thinking. Would the questions select for nerdier but less compassionate candidates? A fair question, but I think not.

WANNA BE PRESIDENT? PASS THIS TEST

A group of well-known scientists calling itself Science Debate 2008 famously called for a presidential debate on scientific issues. Such a forum was a most-welcome development, but I would supplement it with something more revealing of mental firepower: garden-variety puzzles.

Big high-tech corporations such as Google and Microsoft as well as a host of smaller ones routinely utilize puzzles in their hiring practices. The rationale for this is the belief that an employee, say, a programmer of some sort, is more likely to contribute in a creative, insightful way to the company if they're creative and insightful when presented with a complex puzzle.

Why, then, are candidates for the presidency never presented with a few simple puzzles to help the electorate gauge their cognitive agility? The same goes for interviewers who ask the same dreary, insipid questions time after time and accept the same dreary, insipid nonanswers time after time.

These puzzles shouldn't be difficult since, after all, the primary job of the president is to enforce the Constitution, ensure an honest and open administration, and, in some generalized sense, make things better. For this task, judgment and wisdom are more essential than the ability to solve puzzles. Nevertheless, I think some nonstandard questions like the following would help winnow (or at least chasten) some of the candidates. Your guess about how well most aspirants will do with the five questions below?

The Puzzles

1. Scaling. Imagine a small state or city with, let's say, a million people and an imaginative and efficient health care program. The program is not necessarily going to work in a vast country with a population that is 300 times as large. Similarly, a flourishing small company that expands rapidly often becomes an unwieldy large one. Problems and surprises arise as we move from the small to the large since social phenomena generally do not scale upward in a regular or proportional manner.

A simple yet abstract problem of this type? How about the following (answers below): A model car, an exact replica of a real one in scale, weight, material, and so on, is 6 inches (1/2 foot) long, and the real car is 15 feet long, 30 times as long. If the circumference of a wheel on the model is 3 inches, what is the circumference of a wheel on the real car? If the hood of the model car has an area of 4 square inches, what is the area of the real car's hood? And if the model car weighs 4 pounds, what does the real car weigh?

2. Estimating. Proposing any sort of legislation or any action at all requires at least a rough estimate of quantity, costs, benefits, and other effects. An ability to gauge them is crucial (as is an ability to listen to others' unbiased estimates).

A couple of simple yet abstract problems of this type? How about the following: A classic problem: Approximately how many piano tuners are there in New York City? And how many multiples of the death toll on 9-11 is the annual highway death toll?

3. Sequencing. A president must think about how to gain support for an idea or policy. Some things must be accomplished before other things can be attempted. Legislative backing, popular opinion, and domestic and international issues must be dealt with in a reasonable order if an administration is going to be successful. Steps can't be skipped with impunity.

A simple yet abstract problem of this type? How about the following: It's very dark, and four mountain climbers stand before a very rickety rope bridge that spans a wide chasm. They know that the bridge can safely hold only two people and that they possess only one flashlight, which is needed to avoid the holes in the bridge. For various reasons, one of the hikers can cross the bridge in 1 minute, another in 2 minutes, a third in 5 minutes, and the fourth, who's been injured, in 10 minutes. Alas, when two people walk across the bridge, they can go only as fast as the slower of the two hikers. How can they all cross the bridge in 17 minutes?

4. Calculation. Being able to solve a problem using a bit of algebra, it should go without saying, can be useful to a politician, whether the issue is taxes, health policy, or stockbroker commissions.

A simple yet abstract problem of this type? How about the following, which is not irrelevant to broker commissions: A 100-pound sack of potatoes is 99% water by weight. After staying outdoors for a while, it is found to be only 98% water. How much does it weigh now?

5. Deduction. Again, it should go without saying that the ability to make simple deductions is a prerequisite for good decision making.

A simple yet abstract problem of this type? How about the following: Imagine there are three closed boxes, each full of marbles on a table before you. They're labeled "all blue marbles," "all red marbles," and "blue and red marbles." You're told that the labels do describe the contents of the boxes, but all three labels are pasted on the wrong boxes. You may reach into only one box blindfolded and remove only one marble. Which box should you select from to enable you to correctly label the boxes?

Although these problems are much easier than those employers use when hiring entry-level programmers, it would be nice to know that the various candidates, who often are more given to bombast than to logic or evidence, could solve them with ease (although being able to solve them wouldn't be a guarantee of anything). The venue for their solution would be a quiet study with no aides, no pundits, no hot lights, and no intense scrutiny.

What's your guess about how the various candidates would fare with such puzzles? Mine is that a few would find most of the problems trivial, some would have difficulty with them, and the rest wouldn't be sufficiently patient to even try them.

The Solutions:

Answers to 1.): 90 inches, 3,600 square inches, 108,000 pounds. (The area increases by a factor of 30^2, or 900, and the volume or weight increases by a factor of 30^3, or 27,000.)

Hint and answer to 2.): Estimate the population of New York City, the number of households in the city, the percentage of them (and other organizations) that have pianos, how frequently each piano will be tuned on average, and how many pianos the average tuner tunes and put these together for a rough estimate. (Such problems are called Fermi problems in honor of the Italian physicist Enrico Fermi.) The number of deaths in the 9-11 attacks was about one-twelfth the annual highway toll.

Answer to 3.): Label the hikers with their times. First 1 and 2 go over (2 minutes), and 1 comes back (1 minute). Then 5 and 10 go over (10 minutes), and 2 returns (2 minutes). Finally, 1 and 2 go over (2 minutes). The total is 17 minutes.

Answer to 4.): Since the 100-pound sack of potatoes was 99% water, it consisted of 99 pounds of water and 1 pound of pure potato essence. After the evaporation, the sack weighed X pounds and was 98% water and 2% potato essence. Thus, 2% of the new weight X is the 1 pound of potato essence. Since .02X = 1, we can solve to get that X = 50 pounds. The answer is that the potatoes now weigh just 50 pounds. This may seem an apolitical problem, but imagine your stockbroker's fixed fee constituting 1% of the original worth of your investment but 2% of its present worth. Then the problem is not necessarily small potatoes.

Answer to 5.): You would take one marble from the box labeled "blue and red." Assume it's red. (Analogous reasoning holds if it's blue.) Since the marble is red and it comes from an incorrectly labeled box saying "blue and red," it must be the box with red marbles only. Thus, the box labeled "blue" must have either red marbles only or red and blue marbles. It can't be the box with the red marbles only, so it must be the box with blue and red marbles. Finally, the box labeled "red" must contain the blue marbles.

The number pi gets all the publicity and even gets its own day, March 14 (3.14159 . . .). Perhaps another important number, "e," should get its day, February 7 (2.71828 . . .). Let me instead imagine how this number e, which pervades so much of mathematics, might fit into several best-selling plots. It would also be nice if tee shirts (or e shirts) featuring some aspect of the number e would be made. I have three such tee shirts featuring pi, one of which says that my passwords are the last 8 digits of pi. (A couple of people have asked if it's wise of me to reveal my passwords. I assure them that I'm okay with it.)

NOW FEATURING E: PI HAS LONG BEEN IN THE SPOTLIGHT. WHAT ABOUT E?

Specific numbers sometimes play a role in fiction. Witness the novel *The Da Vinci Code*, where the number is the Golden Ratio symbolized by the Greek letter phi, or the movie *Pi*, where the number is pi, of course.

The Da Vinci Code, a thriller offering an alternative view of various conundrums in Western history ranging from the Holy Grail to Mona Lisa's smile, is dependent on the decoding power of phi and the Fibonacci numbers. *Pi* is about a numerologically obsessed mathematician who thinks he's found the secret to just about everything in the decimal expansion of pi and is pursued by religious zealots, greedy financiers, and others. Reflecting on the use of these numbers in fiction, I wondered how a number that doesn't get as much attention as phi or pi might serve as a plot element in a mystery. The number does not have a Greek name but must make do with a simple moniker: e. The base of the natural logarithm and truly one of the most important numbers in all of mathematics, e is approximately 2.71828182845904 (approximately because its decimal expansion continues without repetition).

The first part of an e-based story might briefly sketch the theoretical importance of e and its role in finance, number theory, physics, geometry, and so on. The number might then pop up inexplicably. Here are some possibilities.

Four Mysterious Appearances of e

I. A thriller about outer space. Physicist Robert Matthews has written that, looked at in the right way, the night sky contains the signature of the number pi. Looked at in a different way, the sky also reflects the number e.

Here's the mathematical telescope that allows us to see it: Divide up a square portion of the night sky into a very large number, N, of equal smaller squares. That is, imagine a celestial checkerboard. Then search for the N brightest stars in this portion of the sky and count how many of the N smaller squares contain none of these N brightest stars. Call this number U. (We're assuming the stars are distributed randomly, so by chance some of the smaller squares will contain one or more of the brightest stars, others none.)

If one knows some probability theory, it's not hard to prove that the ratio of N to U (N divided by U, that is) is very close to e and approaches it more and more closely as N gets large. If one doesn't know probability, the appearance of the ratio could seem quite portentous. Before trying to come up with a plot twist that links the number e, this celestial map of the night sky, and some cosmic event, check out the claim. Find a regular 8 × 8 checkerboard, a random number generator, and 64 checkers placed randomly according to the dictates of the generator.

II. A gambling mystery. How might e arise in such a story? A somewhat unusual appearance of the number involves two decks of cards. Shuffle each deck thoroughly, turn over a card from each, and note if it's the same card (both 7s of diamonds, for example, or both jacks of clubs). Then turn over another card from each deck and note if it's the same card. Continue doing this until all 52 cards in the decks are turned over. It can be shown that the probability of no matches at all between the two decks during this sequence of turnovers is extremely close to one chance in e; that is, the probability is 1/e, or about 37%.

Equivalently, about 63% of the time there will be at least one match between the two decks sometime during this turnover process. Again, try it yourself and then figure out a plot element that depends on this appearance of e and the surprising frequency of matches during this process.

III. The number e might also pop up when we are interested in record-breaking events. To illustrate, imagine this year's high school graduates running a quarter-mile race. Runners are randomly selected, and sequentially during a period of months, each of them runs a quarter mile, and we keep track of the number of record times that they establish. The first runner would surely establish a record time, and perhaps the fourth runner would be faster than the first three and establish the second record time. We might then have to wait until the 17th runner, who runs faster than each of the previous 16 runners to establish the third record time. If we were to continue recording times for, say, 10,000 runners, we would find that there would have been only about nine record times.

If we were to keep measuring the times of 1 million runners, we would probably note only about 14 record times. It is no coincidence that the 9th root of 10,000 and the 14th root of 1 million are approximately equal to e. If the Nth runner sets the Rth record, it can be proved that the Rth root of N will be an approximation to e, and this approximation approaches e more and more closely as N increases without bound. This is harder to verify empirically than the previous examples, but you can try.

IV. Idly picking numbers at random can also give rise to e. Using a calculator, pick a random whole number between 1 and 1,000. (A bit better: Pick any decimal number between 0 and 1,000. Say you pick 381.) Pick another random number (say, 191) and add it to the first (which, in this case, results in 572). Continue picking random numbers between 1 and 1,000 and adding them to the sum of the previously picked random numbers. Stop only when the sum exceeds 1,000. (If the third number were 613, for example, the sum would exceed 1,000 after three picks.)

How many random numbers, on average, will you need to pick? In other words, if a large group of people went through this procedure, generated numbers between 1 and 1,000, kept adding them until the sum exceeded 1,000, and recorded the number of picks needed, the average number of picks would be, you guessed it, very close to e. One could be excused for thinking that e stood for "everywhere."

The number e plays a crucial role in all of mathematics, and there are many more beautiful, surprising, and cryptic manifestations of the number in everyday situations (including the process of selecting a spouse). A mystery novel about some of them, perhaps titled *E-erie E-ncounters with E-nigma*, might even be a best seller, perhaps with a list price of 10e dollars: $27.18.

Here is another counterintuitive puzzle that can be presented to beginning students of probability. Note that if the mathematical explanation for why it works is not given or not understood, the puzzle could be a basis for a "psychic" hoax. The same general dynamic holds for more elaborate tricks, both mathematical and otherwise. Make the audience (voters) feel that all reasonable explanations have been foreclosed and that only the supernatural (demagogic) one makes sense.

And once again, a puzzle clarifies a real-life issue in a distant cognitive domain: error-correcting codes.

PUT ON YOUR HATS AND CODES: A HAT PUZZLE AND ERROR-CORRECTING CODES

A simple new puzzle has been tormenting mathematicians and computer scientists since Dr. Todd Ebert, a computer science professor at the University of California, Irvine, introduced it in his PhD thesis in 1998.

Taken up more recently by Peter Winkler of Bell Labs and other puzzle mavens throughout the country, it has become the subject of discussion at high-tech companies, science chat rooms, and university math and computer science departments, in part because of its applications to so-called error-correcting codes.

Here's the situation. Three people enter a room sequentially, and a red or a blue hat is placed on each of their heads, depending on whether a coin lands heads or tails. Once in the room, they can see the hat color of each of the other two people but not their own hat color. They can't communicate with each other in any way, but each has the option of guessing the color of his or her own hat. If at least one person guesses right and no one guesses wrong, they'll each win a million dollars. If no one guesses correctly or at least one person guesses wrong, they win nothing.

The three people are allowed to confer about a possible strategy before entering the room, however. They may decide, for example, that only one designated person will guess his own hat color and the other two will remain silent, a strategy that will result in a 50% chance of winning the money. Can they come up with a strategy that works more frequently?

Most observers think that this is impossible because the hat colors are independent of each other and none of the three people can learn anything about his or her hat color by looking at the hat colors of the others. Any guess is as likely to be wrong as right.

Is there a strategy the group can follow that results in its winning the money more than 50% of the time? The solution and a discussion are below, but you might want to think about the problem before reading on.

There is, in fact, a strategy that enables the group to win 75% of the time. It requires each one of the three to inspect the hat colors of the other two and then, if the colors are the same, to guess his or her own hat to be the opposite color. When any one of the three sees that the hat colors of the other two differ, he or she must remain silent and not make a guess.

A listing of all possible hat colors for the three helps us see why this strategy works. The eight possibilities for the hat colors of the three people are *RRR*, **RRB, RBR, BRR, BBR, BRB, RBB,** and *BBB*, the first entry indicating all three wearing red hats, the second indicating the first two wearing red and the third one wearing blue, the third indicating the first and third wearing red and the second wearing blue, and so on.

In six of the eight possibilities, exactly two of the three people have the same color hat. In these six cases, both of these people would remain silent (why?), but the remaining person, seeing the same hat color on the other two, would guess the opposite color for his or her own hat and be right. In two of the eight possibilities, all three have the same-color hat, and so each of the three would guess that their hats were the opposite color, and all three of them would be wrong.

So when they're wrong, they're very wrong, but when they're right, they're just right enough.

If one were to run this game repeatedly, the number of right and wrong guesses would be equal even though the group as a whole would win the money six out of eight times, or 75% of the time. That is, half of all individual guesses are wrong, but three-fourths of the group responses are right!

Generalizations?

Generalizations to situations with more than three people exist, but all solutions depend on finding a strategy that most of the time results in no one being wrong and every once in a while has everyone being wrong. With seven people playing, a strategy can be devised that wins the money 7/8 of the time, with 15 players, 15/16 of the time.

Are there other situations, say, in the stock market, where independent pieces of information are provided to members of a group who can easily become aware of others' information but not of their own and, hence, where such strategies might work?

The idea behind these strategies can be phrased in terms of error-correcting codes, which are used in compact discs, modems, cell phones, and a host of other electronic devices. Not unrelated is the digit at the end of a barcode, a check digit that is (the units digit in) the sum of all the previous digits in the bar code.

P.S. Use As a Hoax?

Finally, it occurs to me that, were I so inclined, I could exploit the puzzle to appeal to gullible people desperate to find "evidence" for psychic phenomena.

After muttering a few incomprehensible New Age platitudes, I could describe the outcomes as a result of the three people attempting to mentally transmit hat colors and could further claim that whoever receives a strong enough signal from the others will speak up. Then I could observe that 3/4 of the groups respond correctly and attribute this to their telepathic abilities.

Furthermore, by instructing my three co-conspirators not to follow the guessing rules strictly, I could further muddy the waters and inspire speculation about why people more often respond correctly when both of the other two are wearing the same color hat.

I'm always amused when, for example, I see that a 15-inch pizza is $20 and a 10-inch pizza is $16 when the former is more than twice the area of the former. In general, scaling issues provide a lesson of which small businesses and local governments are often unaware. The analogy between a mom-and-pop restaurant and a large business or government agency often founders on these issues. "Tightening your belt," for example, may be useful advice for a small diner or a family of four, but isn't very helpful for larger organizations. A military example of the importance of scaling is provided by Lanchester's square law below.

SCALING UP IS SO VERY HARD TO DO

One of the simplest things you can do to a physical object is to make it bigger or smaller. But the consequences of this change in scale are, I've found, not always obvious to many people. What follows are a few questions about the results of changes in the scale of physical objects. I'll also suggest with an example that similar problems occur when changing the scale of social organizations.

Before you, let's assume, is a very realistic model train car that, down to the last detail, is made out of the same material as the real train car but is 1/20 of the dimensions of the real one. A simple set of questions: If the model is 18 inches long, requires 2 pints of paint to be covered, and weighs 9 pounds, how much paint is needed to cover the real train car, and what is its weight? Answer below.

Ordering Pizzas

You go into a pizza place and notice that the standard 14-inch pizza is $12, while the smaller 10-inch one is $9. You're hungry and poor and need as much for your money as possible. Which do you choose? Similarly, what gives you the most meat: 40 small meatballs, each 1 inch in diameter, or two large ones, each 3 inches in diameter? Answer below.

From King Kong's Problem to Lanchester's Square Law

If a normal gorilla is, let's say, 6 feet tall and 350 pounds, why can't there be a King Kong–sized gorilla 10 times the height but proportioned like a normal gorilla? Answer below. Programs and organizations that work quite well at a certain size often break down when scaled up in size. Consider, to provide a preliminary example, an organization with 8 members, each of whom needs to interact with all 7 of the other members.

This is feasible because there are 28 possible ways to pair the 8 individuals into two-person subgroups. What if the group grows by a factor of 5 to 40 members and it's still necessary for all possible pairs to work together? Simple combinatorics shows that the number of possible ways to pair up the 40 individuals into two-person subgroups grows to 780.

Scaling often has more significant implications, even involving military strategy and advantage. Let me illustrate with Lanchester's square law, formulated during World War I and taught in military schools since then.

Imagine a very simplified conflict between two armies, denoted army A and army B, each of which has 400 pieces of artillery. Assume furthermore that the two sides' artilleries are more or less equivalent in effectiveness and are capable of destroying the other's artillery at a rate of X% of the total per day.

Neither side has an advantage, but let's alter the balance of power and assume that army A can increase its artillery to 1,200 pieces, three times as many as army B has, and that everything else remains the same.

There are two consequences. One is that each of B's artillery pieces will take three times as much fire from A's artillery as before because A now has three times as many guns as B. Because of this, B will lose artillery at three times its previous rate.

The other consequence is that each piece of A's artillery will take only one-third as much fire from B's artillery as before because B now has only one-third as many guns as A. Because of this, A will lose its artillery at 1/3 of its previous rate.

In this case, Lanchester's square law states that tripling the number of pieces of army A's artillery leads to a ninefold advantage in its relative effectiveness, all things being equal. If because of better technology the *quality* of B's artillery improved by a factor of 9, this would equalize things once again. In general, it takes an N^2-fold increase in quality to make up for an N-fold increase in quantity.

Implications of scaling laws in biology are especially important, extensive, and revealing. This is surprising considering how simple such laws are and how complex biological entities are.

SCALING IN BIOLOGY, METABOLIC RATES, AND A PUZZLE ABOUT EVOLUTIONARY TIME

Fascinating new scientific papers suggest how elementary geometry involving animals' physical dimensions is sufficient to shed light on some very basic biological phenomena. In particular, the papers attempt to determine the metabolic pace of all life and, in the process, help resolve a problem in evolutionary time measurement. Hold on tight.

Let's start by considering why an animal can't be, say, 5 times its normal adult size. To understand that we can't simply multiply physical dimensions by a factor of 5, imagine what would happen if a 6-foot, 160-pound man were scaled up to a height of 30 (6 × 5) feet. His weight, like his volume (in cubic feet), would increase not by a factor of 5 but by a factor of 5^3 and thus would rise from 160 pounds to 20,000 pounds (160 × 5^3)—125 times as great as his original weight if he were proportioned similarly.

And what would hold up such a behemoth? The supporting cross-sectional area of his thighs, say, 2 square feet originally, would increase not by a factor of 5 but by a factor of 5^2 and would thus rise to 50 square feet (2 × 5^2)—25 times as great as the normal area if he were proportioned similarly. (The same would hold for his spine, knees, and so on.) But the pressure on his thighs—his weight divided by the area of a cross section of his thighs, that is, 125 times his original weight divided by 25 times the original area—would be 5 times as great. This pressure would be crushing, and the man would collapse. This is why heavy land animals like elephants and rhinos have such thick legs.

Metabolic Rates: Live Fast and Die

Mathematical considerations not too dissimilar to these also lie behind various scaling laws in biology relating animals' metabolic rates—heart, breathing, twitching, and so on—to their surface areas and masses. Small animals' hearts, for example, beat faster than large animals' hearts, and, more generally, they live faster and die younger than do large animals that measure out their energies at a more lumbering pace.

Since areas, including animals' surface areas, scale up with the second power of their relative dimensions, and their masses or volumes scale up with the third power, such considerations long ago led scientists to the belief that animals' metabolic rates were proportional to the surface areas of their skins or, equivalently, proportional to their masses to the 2/3rd power. Much evidence suggests, however, that metabolic rates are proportional to animals' masses to the 3/4th power, not the 2/3rd power, and recent papers by ecologists Brian Enquist and James Brown and physicist Geoffrey West explain why this and other "quarter power scaling laws" make sense theoretically as well as empirically. (The short explanation is that the metabolic rate is affected not by how fast heat dissipates through the skin but rather by how efficiently nutrients reach the body's cells, and for this latter quantity, a broader definition of "surface" area reflecting internal structure is needed.)

In the July 2004 issue of *Ecology*, Enquist, Brown, and West go further and suggest that their quarter power formula (with some refinements involving not only mass but also temperature) describes the metabolic rates of all living organisms, plant as well as animal. As an associate, biologist James Gillooly, has observed, when you correct for size and temperature, the metabolic rates of a shark, a tomato plant, and a tree are remarkably similar. The authors even claim that these metabolic considerations apply generally to biological phenomena on all scales, ranging from the mutation rate for DNA to the speed at which ecosystems change. Again, very simplistically put, small, hot phenomena proceed at a faster pace than do large, cold ones.

Resolution of a Puzzle Involving Evolutionary Divergence

At the risk of losing readers who refuse to attend IMAX theaters showing evolution-themed work, I note that it is this line of thought that also leads to a nice resolution of a problem in evolution. Why do various methods of calculating when two species branched apart often lead to quite different answers?

For example, by examining the DNA of rats and mice, seeing how many dissimilarities there are, and calculating how long it would take for this many mutations to come about, geneticists have placed the branching at around 40 million years ago. But archaeologists looking at the fossils say that the divergence between rats and mice occurred much more recently, about 12 million years ago.

How can we reconcile these numbers? One answer suggested by Gil-looly is based on the work above on metabolic rates. In the January 2005 issue of the *Proceedings of the National Academy of Sciences*, he and his colleagues stress that although small animals don't live as long as larger ones, their metabolic rates are such that their life spans, when measured by these rates rather than by physical times, are comparable to those of bigger animals.

In other words, rats and mice, being small, live at a faster metabolic pace than do larger animals. Because of this and an associated quicker accumulation of mutation-inducing free radicals, their DNA mutates faster than that of larger animals, and hence they require less physical time to diverge as much as they have, not 40 million or so years but approximately 12 million years as the fossil record indicates. Similar reconciliations using animal-specific "metabolic clocks" rather than physical ones exist for other pairs of small animals.

Of course, these scaling laws are crude measuring instruments and admit of many exceptions, and not every biologist is convinced of their utility. Nevertheless, scaling laws do give us a rough handle on metabolic rates that is on the whole very suggestive. They also explain why there are no 30-foot-tall people walking around, except possibly in the imagination of those who believe that humans and dinosaurs co-existed.

Parity refers to the evenness or oddness of a number. Once again, it's a bit surprising that such a simple notion can sometimes be revealing. Furthermore, the revelation generally requires almost no calculation, which is often a characteristic of the best sort of mathematics.

An introductory example: Imagine a 3 × 3 × 3 wooden cube before you, consisting of 27 little cubes. These little cubes are colored red or green, but no two adjacent cubes (having a shared face) are the same color. A termite is introduced to one of the little corner cubes and is instructed to eat through all of the cubes by moving either left or right, up or down, forward or backward, and further directed to die in the innermost cube. Why can't the termite accomplish this?

FIVE OR SIX REASONS WHY PARITY PUZZLES ARE FUN

In recognition of April being Math Awareness Month, my column this month will deal with parity.

The notion refers to the evenness or oddness of a number, say, April, the fourth month, versus May, the fifth. Despite its simplicity, parity plays an important role in many areas of mathematics.

It also lends itself to some nice little puzzles, including Rubik's cube and the 15 puzzle. Here are five or six easy examples. The sixth one is fuzzy and involves politics and the Supreme Court, so it doesn't really count.

The answers to the puzzles appear at the end of the column, but don't peek first—unless, of course, you feel like peeking.

1. A loose-leaf notebook consists of 100 sheets of paper. Number them, front and back, from 1 to 200. Tear out any 25 of the sheets and add up the 50 page numbers on them. Can you choose the sheets so that the sum of the 50 numbers is 2,010?

2. Consider the sum of the first 10 numbers: $1 + 2 + 3 + 4 + 5 + 6 + 7 + 8 + 9 + 10$. Can you change some of the plus signs to minus signs so that the resulting sum is 0? For example, $1 + 2 - 3 - 4 - 5 - 6 + 7 - 8 + 9 + 10 = 3$. This is close but not 0.

3. Before you is a regular 8×8 checkerboard with two diagonally opposite squares missing. Also before you are 31 dominoes, each 2 squares long, 1 square wide. Since each domino covers 2 squares when placed on the checkerboard, 31 dominoes should be enough to cover this mutilated checkerboard. (A regular board has $8 \times 8 = 64$ squares, so this one has $64 - 2 = 62$ squares, and $2 \times 31 = 62$.) So, can you cover this mutilated checkerboard with the 31 dominoes?

4. Two equally matched teams, A and B, play in a best-of-seven World Series. The probability that team A (or team B) wins any given game is 50%, and the first team to win four games wins the series. The series ends at that point. Is it more likely that the series will end in six games or seven?

5. A dozen prisoners are told that they'll be lined up in the morning facing a wall and each looking at the backs of those prisoners ahead of them in line. They're also told that either a red hat or a blue hat will be placed on each of them. They won't be able to see the color of their own hat but will be able to see the color of the hats of those in front of them. After lining up the prisoners in this way, the guards will go to the last person in line and ask him what color his hat is. If he gives the correct color, he will be released, but if he answers incorrectly, he will be killed. Then the next-to-last person will be asked the color of his hat and released if he answers correctly and killed if not and so on up the line. The prisoners are told of this impending procedure the night before and try to decide on a strategy that will save as many of their lives as possible. What should they decide?

6. The Supreme Court has nine justices. The majority rules, and on many issues, the court splits 5–4, with the five more conservative justices prevailing. President Obama will decide soon on whom he will nominate to replace Justice Stevens, one of the four more liberal

justices. What is his best shot at changing the court and its rulings in the short term?

Answers

1. The sum of the page numbers on opposite sides of a sheet of paper must be odd because the pages are numbered consecutively, making one of them odd and the other even and thus their sum odd. If we add up 25 such odd numbers, we'll always get an odd number, but 2,010 is even.

2. First notice that the sum of the first 10 numbers is 55. That is, $1 + 2 + 3 + 4 + 5 + 6 + 7 + 8 + 9 + 10 = 55$. Consider now what happens if we change +3 to –3. The sum is decreased by 6. If we change +8 to –8, the sum is decreased by 16. In fact, whenever we change a sign from + to –, we decrease the sum by an even number. Starting at the odd number of 55 and subtracting even numbers from it will always lead to an odd number, but 0 is even.

3. Since this is a checkerboard, alternate squares are colored red and black, and so the 2 missing diagonally opposite squares are both the same color, say, red. A normal checkerboard has 32 red and 32 black squares, but this one has only 30 red and 32 black squares. This makes covering it with 31 dominoes impossible since each domino will cover 1 red and 1 black square.

4. For the series to end in either six or seven games, it is necessary for it to go more than five games. Thus, after five games, one of the teams, say, it's A, is ahead three games to two. (If one of the teams were ahead four games to one, the series would be over after five games.) So it's three games to two in favor of A. If A wins, the series lasts exactly six games, and if B wins, the series goes into the seventh game. Since each team is equally likely to win, it's equally likely that the series will go six or seven games.

5. The best strategy is for the last person in line (prisoner number 12) to answer red if he sees an even number of red hats in front of him and blue if he sees an odd number. Unfortunately for him, he'll be right

only 50% of the time, but his answer enables all of the other prisoners to survive. If the last prisoner (12) answers red, the prisoner in front of him (number 11) knows that 12 saw an even number of red hats. If 11 also sees an even number of red hats in front of him, he knows that his hat is blue and answers accordingly. If 11 sees an odd number of red hats in front of him, he knows his hat is blue and answers accordingly and so on for prisoner numbers 10, 9, 8, and so on.

6. Since Obama wishes to move the court in a more liberal direction, the only way this might come about in the short term is to pick not necessarily the most eloquent or learned judge but rather that judge who would be most likely to persuade Justice Kennedy, the swing justice among the five more conservative justices. This would switch the court's vote in certain cases from 5–4 to 4–5.

   Simplistic and nowhere near as clear-cut as the straight math problems, this last example is at least tenuously related to the issue of parity. Incidentally, from what I've read about her, Judge Diane Wood would be the most likely replacement for the recently retired Justice Stevens to bring this change to the court.

This is an odd result involving course loads or marketing shares that is related to Simpson's paradox, which is itself rather odd and can lead to serious misapprehensions. (See below.)

COURSE LOADS: AVERAGE PARADOXES THAT WENT TO COLLEGE

Education statistics are a hot topic, and, as usual, they are being spun left and right. Even simple notions like averages can lead to trouble, with odd things happening if one carelessly averages several averages.

My topic for this month concerns two such odd results: one old but still surprising, the other just published.

Sex Discrimination Paradox

First the well-known one, which is often called Simpson's paradox (no, not that Simpson), which has been discussed in numerous popular venues including one of my books. It involves a sex discrimination case in California a while ago that has become a sort of classic illustration. Looking at the proportion of women admitted to the graduate school at the University of California, some women sued the university claiming they were being discriminated against by the graduate school. When administrators looked for which departments were most guilty, however, they were a little astonished to find that there was actually a positive bias for women.

To keep things simple, let's suppose there were only two departments in the graduate school: economics and psychology. Making up numbers, let's further assume that 70 of 100 men (70%) who applied to the economics department were admitted and that 15 of 20 women (75%) were. Assume also that 5 out of 20 men (25%) who applied to the psychology department were admitted and that 35 of 100 women (35%) were. Note that in each department a higher percentage of women was admitted.

If we amalgamate the numbers, however, we see what prompted the lawsuit: 75 of the 120 male applicants (62.5%) were admitted to the graduate school as a whole, whereas only 50 of the 120 female applicants (41.7%) were.

Such odd results can surface in a variety of contexts. For example, a certain medication X may have a higher success rate than another medication Y in several different studies, and yet medication Y may have a higher overall success rate. Or a baseball player may have a lower batting average than another player against left-handed pitchers and also have a lower batting average than the other player against right-handed pitchers but have a higher overall batting average than the other player.

Of course, these counterintuitive results don't usually occur, but they do often enough for us to be wary of uncritically combining numbers and research studies into a so-called meta-analysis.

Course Load Paradox

A new oddity that has a similar flavor was published in the spring issue of *Chance* magazine by Randy Mason, and it concerns a paradox in a common way of presenting marketing statistics (although I'll state it here in terms of college courses taken). Involving overlapping groups of people, the paradox is best understood via two tables.

The first table shows the outcomes of a survey of eight students who were asked about the number and type of college courses they took during the previous year.

The second table gives the average number of courses taken by the eight people and the average number of courses taken by those who took specific types of courses.

Number of courses taken of type:

Student	Math	Science	Social Science	Humanities	Total Courses
1	1	5	1	6	13
2	1	0	0	0	1
3	0	1	0	0	1
4	0	0	2	0	2
5	0	0	0	1	1
6	4	1	6	0	11
7	0	8	0	0	8
8	0	0	0	9	9
					46

Average number of courses taken:

Number of courses taken during last year	Among those who took any course	Among those who took a math course	Among those who took a science course	Among those who took a social science course	Among those who took a humanities course
	46	25	33	26	23
Number of students	8	3	4	3	3
Average number taken	46/8 = 5.75	25/3 = 8.33	33/4 = 8.25	26/3 = 8.67	23/3 = 7.67

A possible problem arises when we note that the average number of courses taken by all eight people is smaller than the average number of courses taken by those who took at least one course of a particular type. That is, if we look only at the three people who took at least one math course, we find that the average number of courses they took (8.33) is bigger than the average number of courses taken by all eight people (5.75).

Okay, you say, maybe there's something special about those people who take math courses that explains why they take more courses on average.

But if we look only at the four people who took at least one science course, we find that the average number of courses they took is also bigger (8.25) than the average number of courses taken by all eight people. Again, we might think there's something special about those people who take science courses that explains why they take more courses on average.

The full paradox appears after we finish checking those who took at least one social science or one humanities course and discover that the same phenomenon holds for them. If we look at those who took at least one course of type X, whatever X is, we will find that the average number of courses taken by this group is higher than the average number reported by the whole group.

It seems that using this particular method of reporting people's course loads can sometimes justify a Lake Wobegon–style boast that those who take any particular type of course take an above-average number of courses.

This odd result may be of use to spin doctors in politics or education. How?

Continuing with the parade of apparent paradoxes, I offer Juan Parrando's example here, in which he describes a game in which two losing strategies can sometimes result in a winning combination.

PARRANDO'S PARADOX: LOSING PLUS LOSING EQUALS WINNING

There's an old story about a store owner who loses money on each individual sale but somehow makes it up in volume of sales.

New calculations by a Spanish physicist now suggest that this paradox may have a kernel of truth to it. Not only does the discovery by Juan Parrondo offer new brain candy for mathematicians, but variations of it may also hold implications for investing strategies.

Parrondo's paradox deals with two games, each of which results in steady losses over time. When these games are played in succession in random order, however, the result is a steady gain. Bad bets strung together to produce big winnings—very strange indeed! To understand it, let's switch from a financial to a spatial metaphor.

Imagine you are standing on stair 0, in the middle of a very long staircase with 1,001 stairs numbered from -500 to 500 (-500, -499, -498, ..., -4, -3, -2, -1, 0, 1, 2, 3, 4, ..., 498, 499, 500).

You want to go up rather than down the staircase, and which direction you move depends on the outcome of coin flips. The first game—let's call it game S—is very simple. You flip a coin and move up a stair whenever it comes up heads and down a stair whenever it comes up tails. The coin is slightly biased, however, and comes up heads 49.5% of the time and tails 50.5%.

It's clear that this is not only a boring game but also a losing one. If you played it long enough, you would move up and down for a while, but almost certainly you would reach the bottom of the staircase after a time. (If stair climbing gives you vertigo, you can substitute winning a dollar for going up a stair and losing one for going down a stair.)

A More Complex Game

The second game—let's continue to wax poetic and call it game C— is more complicated, so bear with me. It involves two coins, one of which, the bad one, comes up heads only 9.5% of the time, tails 90.5%. The other coin, the good one, comes up heads 74.5% of the time, tails 25.5%. As in game S, you move up a stair if the coin you flip comes up heads and move down one if it comes up tails.

But which coin do you flip? If the number of the stair you're on at the time you play game C is a multiple of 3 (i.e., . . . , –9, –6, –3, 0, 3, 6, 9, 12, . . .), then you flip the bad coin. If the number of the stair you're on at the time you play game C is not a multiple of 3, then you flip the good coin. (Note: changing these odd percentages and constraints may affect the game's outcome.)

Let's go through game C's dance steps. If you were on stair number 5, you would flip the good coin to determine your direction, whereas if you were on stair number 6, you would flip the bad coin. The same holds for the negatively numbered stairs. If you were on stair number –2 and playing game C, you would flip the good coin, whereas if you were on stair number –9, you would flip the bad coin.

It's not as clear as it is in game S, but game C is also a losing game. If you played it long enough, you would move up and down for a while, but you almost certainly would reach the bottom of the staircase after a time.

Game C is a losing game because the number of the stair you're on is a multiple of three more often than a third of the time, and thus you must flip the bad coin more often than a third of the time.

So far, so what? Game S is simple and results in steady movement down the staircase to the bottom, and game C is complicated and also results in steady movement down the staircase to the bottom. The fascinating discovery of Parrondo is that if you play these two games in succession in random order (keeping your place on the staircase as you switch between games), you will steadily ascend to the top of the staircase.

So-called Markov chains are needed for a fuller analysis.

Connection to Dot-Com Valuations?

Alternatively, if you play two games of S followed by two games of C followed by two games of S and so on, all the while keeping your place on the staircase as you switch between games, you will also steadily rise to the top of the staircase. (You might want to look up M. C. Escher's paradoxical drawing, *Ascending and Descending*, for a nice visual analogue to Parrondo's paradox.)

Standard stock market investments cannot be modeled by games of this type, but variations of these games might conceivably give rise to counterintuitive investment strategies. Although a much more complex phenomenon, the ever-increasing valuations of some dot-coms with continuous losses may not be as absurd as they seem. Perhaps they'll one day be referred to as Parrondo profits.

Chapter Two

A Bit of Probability

Nothing amuses more harmlessly than computation, and nothing is oftener applicable to real business or speculative enquiries.
—Samuel Johnson

PROBABILITY IS FULL OF NUANCES, SOME OF THEM SIMPLY VERBAL BUT still capable of causing confusion. Flip a coin 10,000 times. The most likely number of heads that will turn up is 5,000, but it is quite unlikely that the number of heads that turn up is 5,000. A most likely outcome, of course, needn't be a likely outcome.

Having mentioned verbal nuances, I note the obvious, that probability and statistics in their formal incarnations often appear somewhat forbidding to people, but I claim the perhaps less obvious, that the subject's technical notions grew out of understandings everyone possesses. Consider the terms for central tendency, for example, mean, median, mode, and so on. These notions grew out of workaday words like "usual," "customary," "typical," "same," "middling," "most," "standard," "stereotypical," "expected," "nondescript," "normal," "ordinary," "conventional," "commonplace," and so on.

Or examine the precursors of notions of statistical variation: standard deviation, variance, and the like. They grew out of and are refinements of words such as "unusual," "peculiar," "strange," "singular," "original," "extreme," "special," "unlike," "unique," "deviant," "dissimilar," "disparate," "different," "bizarre," "too much," and so on.

Probability itself is presaged in everyday words like "chance," "likelihood," "fate," "odds," "gods," "fortune," "luck," "happenstance," "random," and many others.

And consider the essential act of changing our mind about something. We all change our minds about things when we get new information, and we often do it in a grossly inaccurate manner. Bayes' theorem, named after the 18th-century minister Thomas Bayes, is a refinement of this process. The theorem provides us a formal and precise way to update our estimates of the probability of an event in light of new knowledge.

A simple example: There are two coins before us, one fair, one two-headed. We choose one of them and initially assume the probability to be 1/2 that it is the fair one. Then we flip it three times, and it comes up heads all three times. What should we estimate the probability of its being the fair one now? Answer is 1/9.

The bottom line is that probability and statistics are not alien ideas but are in dire need of distillation and clarification. Many of us are like the Molière character who was surprised he spoke prose his whole life. We've all been speaking probability our whole lives but usually quite badly. Our vocabulary for it is sometimes not much more nuanced than "one in a million," "fifty-fifty," and "sure thing." We don't recognize how contingent events are, we don't compare the probabilities of events, we often don't know what probabilities are relevant in a given situation, and we're too impressed by confidently stated certainties.

John Gay's maxim "Lest men suspect your tale untrue, Keep probability in view" is wise, but it makes sense only if we understand, for example, that the probability of hospitalization or death resulting from the COVID vaccine is very much lower than the probability of hospitalization or death resulting from a COVID infection and if we understand that the probability of a particular event happening, say, meeting a particular acquaintance in a mall in Thailand, is very much lower than the probability of an event of that general sort happening, meeting someone you know in an unusual locale. Moreover, of course, we should understand that though "very much lower" and "very much higher" are essential assessments of relative probabilities, they can often be made somewhat more precise.

In any case, regularly asking oneself the question "Roughly how likely is that?" is good cognitive hygiene.

This is a mini-tutorial on independent events and their applications, of which there are countlessly many. One recent specific example, which I'll come to shortly, is the savings possible when pooling blood samples to test for a disease such as COVID.

There are a few reasons why many people find thinking probabilistically so uncongenial if not downright unpleasant.

PROBABILITY AND INDEPENDENT EVENTS: A MINI-TUTORIAL

Probability is a difficult subject for a number of reasons. Perhaps the primary one is that most people don't naturally think in probabilistic terms, at least in a careful way. Their vocabulary for dealing with the subject is meager at best. An event or outcome might be deemed either a sure thing, or 50-50, or else impossible, a bit like the "1, 2, many" numbering system in some Australian Aboriginal languages. This reminds me of Andy Rooney's quote, "The 50-50-90 rule: anytime you have a 50-50 chance of getting something right, there's a 90% probability you'll get it wrong."

I'm also reminded of a conversation I once had with a taxi driver who justified spending $100 on the lottery by explaining that he might win or he might lose, but better a 50-50 chance than no chance at all. One more example among 7 gazillion possible ones is provided by the conventional assessment that the polls in the 2016 election were way off. The probability that Hillary Clinton would win was estimated to be 85%, which many, I'm sure, thought was essentially a sure thing. But a 15% chance of a Clinton loss was about the same as rolling a pair of dice and having a 7 turn up or, equivalently, having a 10, an 11, or a 12 turn up. Not that unlikely at all.

A second reason probability is avoided if not scorned is that many people subscribe to the belief that "everything happens for a reason" and that therefore outcomes are certain to happen or certain not to happen—the "every blade of grass" folks. They lack or have no need for the degree of abstraction necessary to rise above everyday folk psychology or religious doctrine when confronting the world. (At the other end of the abstractness spectrum, there are, of course, a million and seven arguments and discussions on philosophical determinism and quantum theory, but this is not the place for the complications raised by issues as diverse as soft versus hard determinism, action at a distance, quantum entanglement, and Bell's theorem.)

A third obstacle for many is that much of the reasoning and many of the conclusions in probability are tricky and counterintuitive. I'll get to one of the most well known of these conclusions below as well as to a common confusion that is relevant to sentencing protocols in the criminal justice system, but first a little necessary background.

Diving in, let me note that when assessing the probabilities, we must be careful to define and distinguish between two important classes of events: independent events and mutually exclusive ones. Two events are said to be independent if the occurrence of one of them has no effect on the probability of the other occurring. That is, the outcome of one of the events does not affect the (probability of the) outcome of the other one. Two flips of a coin provide a simple illustration. Whatever the outcome of the first flip, the probability that the second flip results in a head is 1/2, as is the probability that it results in a tail. The outcome of the second flip is independent of the outcome of the first. The definition of independence extends as well to three or more events.

What is particularly pleasant about such events is that the probability of both (or all) events occurring is simply the product of each of their individual probabilities. Thus, the probability of flipping two heads in a row is, since the flips are independent, 1/2 × 1/2, or 1/4. Likewise, the probability of flipping three heads in a row is 1/2 × 1/2 × 1/2, or 1/8, which is the same as flipping three tails in a row, which in turn is the same as getting the sequence of tails, heads, tails. No matter what sort of events we consider, if they are independent, we can easily find the probability of all of them occurring.

Thus, the probability of flipping a coin twice and getting heads both times, then rolling a single die three times and getting a 1 all three times, then picking an ace from a 52-card deck, and finally spinning a roulette wheel and having it land on a red sector is simply $(1/2)^2 \times (1/6)^3 \times 4/52 \times (18/38)$, or 18/426,816. Likewise, the probability of surviving three trials of Russian roulette is for the same reason equal to $(5/6)^3$, 125/216, or about 58%. (Russian roulette is the "game" of loading a bullet into one of the six chambers of a revolver, spinning the cylinder to randomize the position of the bullet, and then pulling the trigger while pointing the gun at one's own head.)

The relevance of independence to criminal trials is worth examining. Say a prosecutor presents four independent pieces of evidence (the suspect was near the crime scene, had large feet, wore a beard, and seemed to know a couple of the victim's acquaintances): independent facts that are true of the real murderer and possibly of the suspect on trial. The evidence is circumstantial, and each piece is true of 50%, 40%, 30%, and 20% of the relevant population. The prosecutor concludes that the probability that an innocent suspect would have all four of these pieces of evidence indicating his guilt arrayed against him is, assuming independence, the product of their respective probabilities, $.2 \times .3 \times .4 \times .5$, which equals .012, or 1.2%. The prosecutor then concludes with a flourish that the probability that the suspect is guilty is 98.8%.

This argument is often called the prosecutor's fallacy, and a counter to it will be discussed in a later section. Unfortunately, it often leads to the conviction of innocent people since lawyers, judges, and juries are averse to probabilistic arguments.

If we pool blood samples to save on the number of tests needed for COVID or any other condition, independence enables us to quantify the savings. Let's illustrate with a condition that afflicts 1% of people. Assume now that we pool the blood of 20 randomly selected people. The probability that a person does not have the condition in question is 99%, or .99, so the probability that none of the 20 people have it is, by independence, $.99^{20}$, which equals about .82, or 82%. So 82% of the time, only one test is needed. But 18% of the time, there will be at least one of the 20 people who has the condition, so 18% of the time, 21 tests will be needed: the pooled sample plus 21 individual tests. The average number of tests needed for any group of 20 people is thus $.82 \times 1 + .18 \times 21$, which turns out to be about 4.6 tests rather than 20. A sensible policy indeed, where possible.

A simple exercise: What is more probable: flipping a coin 10 times and getting 10 straight heads or rolling a single die four times and getting four straight 6s?

Finally, I should distinguish here between independent events and mutually exclusive events. The latter are simply events with the property that if any one of them occurs, the others cannot occur. If these events are actions, you can't perform more than one of them. At 8 o'clock, for example, you may go out to dinner, you may go to a ball game, you may go to a movie, or you may do none of these things. Because these events exhaust your possible actions and you must do one of them, the probability that one of them occurs is the sum of all their probabilities, which is 100%, or, converting to decimals, 1.00. If the probability that you go out to dinner is 20% and the probability that you go to a movie is 42%, then the probability that you engage in one of these two activities is then simply .20 + .42, which equals .62, or 62%.

Or say you roll a single die. What is the probability that a 1 or a 6 turns up on the die? The outcomes of 1 and 6 are mutually exclusive, and each of them has a probability of 1/6 of turning up, so the probability that one of these two numbers turns up is 1/6 + 1/6, which is 1/3, or 33.33%. Contrast this with the question: If you roll the die eight times, what is the probability that a 1 or a 6 turns up each of the eight times? Here, since the rolls are independent, the answer is $(1/3)^8$, or 1/6,561, or about .015%.

Students sometimes say they don't know when to add and when to multiply probabilities. If you want the probability of two or more mutually exclusive events occurring, you add their respective probabilities, but if you want the probability of independent events all occurring, you multiply their respective probabilities. (Note that if two events are not mutually exclusive but overlap, say, passing math and passing English, we have to be careful. If, for example, in a very large class 60% of the students pass math and 70% pass English, it would make no sense today that 130% of the students passed at least one of these courses. We have to subtract the percentage of students who passed both courses, whatever it is, from 130% since we don't want to count them twice.)

The average or expected value of a procedure or data set is often important. Roughly, it is simply an average where events or numbers occurring more frequently are counted more heavily. In particular, the expected value is important in the following variations on the task of collecting a complete set of anything, like the complete set of baseball cards that my mother threw out or like a (complete) family with a child of each sex.

Collecting a Complete Set: Baseball Cards, Disease, and Denial of Service

Rolling a die until all six numbers turn up, having children until you have at least one of each sex, and collecting the complete set of baseball cards are three examples of a certain sort of multipart task. To complete such a task, it's necessary to perform several different subtasks. If the probabilities of performing these various subtasks are the same, some relatively simple math helps figure out how long or extensive the complete task will be.

To illustrate, let's start with a die. How many times will you have to roll it on average until each of the six numbers appears at least once?

A couple of ideas are needed to answer this question, the first quite plausible.

Assume that the probability of some event or outcome of interest is a number p and that repeated tries to obtain the outcome have no effect on the probability of later tries. Then the average number of tries needed to obtain the outcome is $1/p$.

This may sound complicated, so let's return to the die example. Since rolling a die randomly generates the numbers 1 through 6, the probability of obtaining a particular number, say, 2, is 1/6, so on average, we'll need $1/(1/6)$, or 6, tries to generate a 2. On average, we'll also need 6 tries to generate a 5 and so on.

By contrast, the probability of rolling an odd number, either 1, 3, or 5, is 3/6, so on average, we'll need $1/(3/6)$, or 2, rolls of the die to roll an odd number.

The second idea builds on the first. The first roll of a die necessarily gives us a number not rolled before. After we've rolled this first number (whatever it is), the probability that we roll a different number is 5/6 since we want to roll one of the 5 numbers we haven't yet rolled.

The average number of additional rolls until a second number turns up is thus 1/(5/6), or 6/5. After rolling two different numbers, the probability that we roll a third number different from the first two is 4/6, and so the average number of additional rolls until it turns up is 1/(4/6), or 6/4. Continuing in this way until all 6 numbers are rolled and adding the results gives us (1 + 6/5 + 6/4 + 6/3 + 6/2 + 6/1), or 14.7 rolls on average to obtain all six numbers. Try it a few times.

The same basic calculation answers the other questions mentioned.

A couple plans to continue having children until they have a child of each sex and wonders how many children they'll likely have.

Or a baseball card collector wonders how many pieces of gum he'll probably have to buy to obtain the complete set of 400 cards. The answer for the average number of children is 3. The first child will necessarily be of a sex not obtained before.

After the first child, the probability of a child of the opposite sex is 1/2, so 1/(1/2), or 2, additional children will be needed on average to obtain a complete set (3 = 1 + 1/(1/2) = 1 + 2).

Likewise, the average number of baseball cards needed to obtain the complete set is 1 + 1/(399/400) + 1/(398/400) + 1/(397/400) + . . . + 1/(2/400) + 1(1/400), which equals more than 2,000 (approximately 400 ×ln(400)).

The last term of 1/(1/400), or 400, additional cards reflects the difficulty of getting that last missing card, probably an obscure utility infielder, to complete your collection.

Further Afield

If we relax the assumption of equal probability for different outcomes, we get more practical applications. One odd example arises in so-called denial-of-service attacks in a computer network.

In these, an attacker repeatedly sends packets of partial information (parts of a file, say) to a site to flood it and make it unavailable.

The site reassembles these randomly arriving packets into a complete file when at least one of each of the packets is received.

How long it takes a complete file (or many complete files) to travel from the attacker to the destination site can be computed using roughly the same mathematics as that used in the examples above.

If each of the routers along the path the information packets take marks them electronically, then how long it takes the destination site to receive the complete set of marked packets can be calculated and the attacker's location sometimes inferred from the marked path taken.

Another possible example is that of a disease that requires at least one of a number of different independent, continually occurring assaults on the body. Once all of these have been collected, as it were, the disease develops.

Dice, children, baseball cards, denial-of-service attacks, and disease together testify to the imperialist nature of mathematics. It sends out colonies to almost all disciplines and endeavors.

Joltin' Joe DiMaggio may have benefited from the New York Yankees' relationship with the official scorer. Whether he did or didn't, the probability of independent events gives us some perspective on the issue.

PROBABILITY AND HITTING STREAKS: DOES DIMAGGIO'S DESERVE AN ASTERISK?

October is the month of baseball's World Series, so it is an appropriate time to consider "56*," an article just published in the Canadian magazine *Walrus*. In it, author David Robbeson asks, "Was Joe DiMaggio's fabled fifty-six-game hitting streak the greatest feat in all of sports or did it also benefit from a little help from his friends?"

The article strongly suggests that DiMaggio's legendary record might deserve an asterisk similar to that many have attached to Barry Bonds's breaking of Henry Aaron's home run record.

Robbeson's argument, which has been made before but never, I think, so thoroughly, revolves around one Dan Daniel. Daniel was a baseball writer who had covered the Yankees for a long time, was a personal friend of many of the players, and traveled with the team and submitted his expenses to it. He was also the official home-game scorer for the Yankees.

He decided, among other things, whether any at-bat should be adjudged a hit by the batter or an error by the fielder, yet he was, in Robbeson's words, "as much a PR man as a reporter."

Specifically, Robbeson cites two games in the middle of the streak, the 30th and 31st, when DiMaggio managed just one hit. In each of these games, the hit was suspect and could well have—and perhaps should have—been deemed an error.

The first involved a bad bounce that hit off the shoulder of shortstop Luke Appling after he reached for it. Hits and errors were not immediately recorded on the scoreboard, so, Robbeson writes, some spectators believed the streak had come to an end. Daniel, however, called it a hit. The 31st game of the streak involved a fielding play that was also arguably an error on the part of Appling, who got his glove on the ball but dropped it. Again, Daniel called it a hit.

How could this have happened without arousing more controversy? Robbeson argues that despite the present Olympian status of the streak, at the time, American involvement in World War II was looming, attention to the then-29-game streak and its fluky extension was not intense, and baseball attendance was quite low. Amazingly, the attendance in 22 of the games during the streak was less than 10,000.

There is no clip of the fabled streak by the Yankee Clipper on YouTube to decide the matter, so it will never be conclusively settled. There is, however, a counterargument, well stated by Stephen Jay Gould.

Gould argues that these two, at best, weak hits as well as a couple of others seem out of place in a record set by a mythical hitter like DiMaggio. The reason is that people tend to believe that streaks are a causal consequence of courage and competence and that their lucky extension is somehow an affront to our conception of them. DiMaggio is too great a figure, people unconsciously think, to have his streak depend on such thin threads.

As psychologists Amos Tversky and Daniel Kahneman demonstrated years ago, however, people fervently yet mistakenly believe in hot hands, in clutch hitters, and in coming through under pressure and don't want to think of streaks as simply matters of luck. But luck is sometimes a big part of it, and good hitters benefit more from it than do bad hitters. They will generally hit in longer streaks than bad hitters just as heads-biased coins will result in longer strings of consecutive heads than tails-biased coins will.

In other words, DiMaggio's streak remained intact because of these calls by Daniel, but so what? Some lucky breaks and a dubious call or two are to be expected in a long streak.

Whether the streak was tainted by a biased scorer operating in a different historical context and under much less fierce media scrutiny can't be cleanly judged now. We can, however, understand something of the improbability of DiMaggio's feat by doing a few little calculations.

His lifetime batting average was .325. If, therefore, we assume as a first approximation that he generally got a hit 32.5% of the time he came to bat and hence made an out 67.5% of the time and further assume that he came to bat four times per game, then the chances of his not getting a hit in any given game were approximately, assuming independence, $(.675)^4 = .2076$.

Remember that independence means he got hits in the same way a coin that lands heads 32.5% of the time gets heads. So the probability that DiMaggio would get at least one hit in any given game was $1 - .2076 = .7924$.

Thus, the chances of his getting a hit in any given sequence of 56 consecutive games was $(.7924)^{56} = .000002192$, a minuscule probability indeed.

The number of times in a season that a hitter with a .325 batting average might be able to hit successfully in exactly 56 consecutive games (going hitless before and after the 56-game streak) is also tiny. This number is determined by adding up the ways in which he might hit safely in some string of exactly 56 consecutive games. The probability of streaks of length at least 56 straight games is about five times higher, but DiMaggio hit in only 139 games, so his chances were somewhat less than this multiple of 5.

The conclusion is that such an extraordinary achievement "should not," probabilistically speaking, have yet occurred in the history of baseball. There are many differences and a few similarities between Bonds's record and DiMaggio's. I leave them for readers to ponder.

Taken literally, this common safety directive simply cannot be obeyed in many situations. The so-called combinatorial coefficients help explain why and also lead to an alternative solution to the notorious birthday problem.

HOW MANY WAYS: FROM SOCIAL DISTANCING TO AN ALTERNATIVE SOLUTION TO THE BIRTHDAY PROBLEM

During the coronavirus pandemic, many states restricted social gatherings to 10 people. Even this small number, however, gives rise to 45 possible handshakes, conversations, and interactions of one sort or another. That is, the number of ways of choosing 2 people out of the 10 is (10 × 9)/2, or 45. That is, each of the 10 people can interact with any of the remaining 9, which can happen in 10 times 9 ways, and then we divide by 2 since if A interacts with B, we don't want to count B interacting with A as a different interaction. So even with just 10 people we have 45 possible interactions and thus 45 possible routes of contagion.

But what if allow gatherings of 50 people? The number of possible interactions would then by the same argument be (50 × 49)/2, or 1,225. So allowing only 5 times as many people would result in more than 27 times (1,225/45) as many possible interactions or possible routes of contagion. Observe that allowing gatherings of 50 is far, far riskier than allowing just 10.

And, to press the point, if we wanted to allow 40,000 people to attend a sporting or musical event, this would make nearly 800 million interactions possible, not to mention requiring a stadium more than 30 miles across to accommodate 6-foot spacing between each person and his or her nearest neighbor. Of course, there'd be no way for any significant number of pairs of people to interact. Nevertheless, there'd be many smaller groups whose members would interact.

I should note parenthetically that there are better ways to pack people into an area besides placing them in a lattice of points each 6 feet apart (like street corners if the city blocks were 6 feet long). In fact, there is a considerable body of research about tightly packing circles that is of mathematical interest. Of course, these latter considerations are not of much practical use in social distancing. The number of possible pairs of people gives us rather absurd upper bounds, but it does illustrate that pedestrian concerns can quickly lead to nontrivial mathematics.

To reiterate, the number of two-person interactions in a group of N people is proportional to N^2 (or N-squared if you don't like exponents). More precisely, it is, as the examples above suggest, [N × (N − 1)]/2. So how many ways are there to choose 2 people out of 23? Plugging into the formula, we find it's (23 × 22)/2, or 253. We'll see in a bit why this case is especially interesting.

More generally, the number of three-person interactions in a group of N people is given by [N × (N − 1) × (N − 2)]/(3 × 2 × 1) and so on. This is the formula for the so-called combinatorial coefficients, which give us the number of ways of choosing a fixed number of elements out of a set of N elements.

This brings me to an intriguing counterintuitive result introduced in most courses in probability: the birthday problem. Since we have two ingredients that are sufficient to yield this classic result, here is a brief, slightly nonstandard derivation of it. Clearly, if we gather 366 people, we can be sure with probability 100% that at least two of them have the same birthday. This is because 366 is one more than 365, the number of possible birthdays (for simplicity, let's forget about February 29). But what's not clear and is quite surprising is that if we gather *just 23 people*, the probability that at least two of them have the same birthday is 50%! How could this be?

As the formula mentioned above earlier shows, if we gather a random set of 23 people, there are 253 possible *pairs* of these 23 people since 253 = [(23 × 22)/2]. Furthermore, given any one particular pair of these 253 pairs, say, George and Martha, the probability that the pair have different birthdays is 364/365. This can be seen by noting that while one of the pair, say, George, can have any birthday, the other's (Martha's) chances of sharing it are 364/365. We randomly chose the 23 people so that their birthdays and hence the 253 birthday pairs are independent. So what is the probability that no pair of people from among the 253 possible pairs of people have the same birthday? By independence, we find it to be $(364/365)^{253}$, which is almost exactly 1/2, or 50%. Therefore, the probability that at least one pair of people does share the same birthday is 1 - 1/2, which equals 1/2, or 50%.

If you are on Facebook and you have at least 23 friends, you may want to randomly choose 23 of them and see if any pair of the ones you've chosen have the same birthday. If you randomly choose 30 people (say, U.S. presidents), the probability that at least two have the same birthday rises to 70%. You might try to mimic the above argument to show this, or you might not.

The notion of conditional probability isn't relevant in cases where we have no background knowledge, but cases like this are rare since we usually know something of the facts relevant to a given event. I'm often asked, for example, What is the probability of X? I don't answer and explain that the event X is complex, one of a kind, and dependent to varying degrees on many factors. Sometimes, people are annoyed by this response and insist on being provided the formula that would yield the right answer.

Even knowledge of a single fact can often drastically change the (conditional) probability. As John Maynard Keynes famously said, "When facts change, I change my mind. What do you do?" The probability that a randomly chosen man weighs more than 270 pounds is quite low, whereas the conditional probability that a randomly chosen man weighs more than 270 pounds given that he is taller than 6 feet, 6 inches is considerably larger.

Conditional probabilities are like conditional statements in logic. Just as "if A, then B" is quite different from "if B, then A," "the conditional probability of A given B" is quite different from "the conditional probability of B given A." More examples: The conditional probability that a person is male if we know the person is the CEO of a large company is quite high, whereas the conditional probability that a person is the CEO given the person is a male is very, very low. Likewise, the conditional probability that someone suffers from a serious mental illness given that the person is a serial murderer is high, whereas the conditional probability that a person is a serial murderer given that the person has a serious mental illness is very, very low.

CONDITIONAL PROBABILITY (BECAUSE WE USUALLY KNOW THINGS) AND THE PROSECUTOR'S FALLACY

The distinction between probability and conditional probability is not always clear, and it can sometimes be a matter of medical, political, or legal significance. The probability that a 40-year-old adult has diabetes is different and much lower than the conditional probability that a 40-year-old adult has diabetes given the person is obese. Likewise, the probability that an American voter casts a vote for a Democrat is different and higher than the conditional probability that an American votes Democratic given that the voter is a resident of Oklahoma City and, contrariwise, lower than the conditional probability that an American votes Democratic given that the person is a resident of San Francisco.

An interesting and sometimes life-changing application of the latter distinctions is provided by the aforementioned prosecutor's fallacy. A prosecutor in a criminal case might well argue that the conditional probability of the evidence, E, presented against the accused person given the assumed innocence, I, of the accused is very small (say, just 1.2% as described in the section above), so the accused must be guilty. Say, for example, the accused person was near the crime scene, had a beard as a witness claimed the perpetrator did, had met a couple of people who knew the victim, and had large feet as indicated by a footprint near the crime scene. Again, this would indicate that the conditional probability of the evidence presented given the assumed innocence of the accused is very small.

So far, no problem, but this is the wrong conditional probability. The correct and relevant conditional probability is that the accused is innocent, I, given the evidence, E, presented and not the conditional probability of the evidence, E, presented given the assumed innocence, I, of the accused. You may want to read the previous sentence again. The conditional probability of I given E could be much larger than the conditional probability of E given I.

The prosecutor might well try to elide the significant difference between these two probabilities, but the defense attorney hopefully would stress their difference by, for example, pointing out that the area was densely populated so that possibly hundreds of people were near the crime scene, many with beards and big feet who knew a couple of the victim's acquaintances and that therefore there might a dozen or more people satisfying all four of the bits of circumstantial evidence. Thus, the conditional probability of I given E is much larger than the conditional probability of E given I (which, remember, was a very misleading 1.2%), and the suspect should be acquitted.

Another example: The only evidence at a crime scene is the DNA of the perpetrator, and a person is found with this DNA. Clearly, the conditional probability of this evidence given the assumed innocence of the accused is minuscule, and so the accused is certainly guilty. Not so fast. Even here, the relevant conditional probability of innocence given the evidence might be much higher. The defense might find the twin brother of the accused who was adopted at birth. He would have the same DNA, so even though conditional probability of E given I is minuscule, the conditional probability of I given E would be 1/2 since there are two people who share the same DNA.

Of course, in many such cases, both conditional probabilities are small, and the accused is indeed guilty.

Terms with a precise meaning often have an imprecise colloquial use, and the confusion of the two can sometimes lead to serious misunderstandings. These misinterpretations are especially common in probability, but they occur in mathematics generally. I can't count the number of times people use the word "exponential" when they mean "fast." Happily, people rarely use the word "logarithmic" to mean slow.

The confusion below is considerably weightier.

FOUR TIMES THE ODDS OF CONVICTION NOT FOUR TIMES AS LIKELY AND OTHER DISTINCTIONS

Of course, probability is much more than the source of cute counterintuitive conclusions. Simply misinterpreting its terms can lead to real-world policy implications as well.

Consider, for example, the case of an academic study that published a figure that was ostensibly correct but seriously misleading. In reporting on death sentences in Philadelphia, the study by the Death Penalty Information Center asserted that the odds of blacks convicted of murder receiving a death sentence were four times the odds faced by whites and other defendants similarly convicted. Since then, many accounts of the study in newspapers and on television have transmuted that statement into the starkly inequivalent one that blacks were four times as likely to be sentenced to death as whites. The authors of the study used the technical definition of odds, and since the odds are that most people do not know the difference between the term and probability, people were seriously misled.

The difference is crucial. The odds of an event are defined as the probability that it will occur divided by the probability that it will not occur. For example, let's look at a coin flip. The probability of its landing heads is one-half, or .5, and the probability of its not landing heads is also one-half, or .5. Hence, the odds of the coin landing heads is 1 to 1 (.5 divided by .5). Or what about rolling a die and being interested in the event that it lands on 1, 2, 3, or 4? The probability of this event's occurrence is four-sixths, or about .67, and the probability of its not occurring is two-sixths, or about .33. Hence, the odds of the die landing on one of these five numbers is 2 to 1 (.67 divided by .33). The same distinction holds for more serious situations. For example, if the probability of someone suffering from some disease is .80, or 80%, the odds that one suffers from it are .80/.20, or 4 to 1.

Even more consequential is the relevance of this to murder statistics and racial disparities in the handing out of death penalties. To most readers, the phrase "four times the odds" means, to illustrate with an extreme example, that if 99% of blacks convicted of murder were to receive the death penalty, about 25% of whites similarly convicted would receive the same penalty. Yet when the technical definition of "odds" is used, the meaning is quite different. In this case, if 99% of blacks convicted of murder received the death penalty, then a considerably less unfair 96% of nonblacks similarly convicted would receive the death penalty. Why?

Using the technical definition, we find that the odds in favor of a convicted black murderer's receiving the death penalty are 99 to 1 (99/100 divided by 1/100). The odds in favor of a convicted nonblack murderer's getting the death penalty are 24 to 1 (96/100 divided by 4/100). Thus, since 99 is roughly four times 24, the odds that a convicted black murderer will receive the death penalty are, in this case, approximately four times the odds that a convicted nonblack murderer will receive the same sentence.

By dissecting the phrase "four times the odds," I certainly don't mean to deny that racism exists, that there are very significant racial disparities in sentencing, or that the death penalty is morally wrong. Rather, I mean to deflate the likely-to-be-inferred magnitude of racial disparities in the sentencing for murder and other violent crimes. The difference between 99% and 96%, for example, is much less egregious than that between 99% and 25%. Still, whatever the percentages are (96% and 86% yield the same four times the odds), the raw numbers are quite troubling enough without the easily misinterpreted phrase "four times the odds."

Another common distinction that is often not made is that between the risk of a certain condition and the relative risk of it. The latter is often mentioned in news stories but doesn't mean much if the absolute risk isn't also given in the story. A couple of examples make this clear. If one's absolute risk of contracting condition X is for most people about 8 in a 1,000, then a relative risk of 1.5 for people or who engage in a certain behavior, involving, say, eating bacon regularly, means that their risk of contracting X is 1.5 × 8, or 12, in 1,000. The relative risk for people who exercise regularly might be .75, which means that their risk of contracting X is .75 × 8, or 6, in 1,000.

Even a relative risk of 3 or 4 for a certain behavior could be reasonably ignored if the absolute risk were 2 in a million since this would up the risk to 6 or 8 in 1 million. Risk versus relative risk is thus a variant of the distinction between probability and conditional probability.

Alas, most mistaken numerical conclusions are not esoteric in the least. I once discussed suicide with a student of mine who had recently lost a brother to suicide. He was understandably distraught and maintained he'd read that there was a suicide every 45 seconds in the United States. Obviously, it wasn't the time to do some long division, but that frequency would be implausibly high. A suicide every 45 seconds would result in about 700,000 suicides annually, whereas it's estimated there are around 50,000 annually from all causes.

It's a nice exercise when an advertisement for a product states that a case of X occurs every Y seconds to do the simple arithmetic needed to determine the number of cases of X annually. I often find it to be a quite dubious total, whether because of innumeracy or a tendentious intent.

Randonauts is an app that uses "quantum-generated numbers" to formalize and have fun with coincidences. It's harmless play as long as you don't assume that they have a deep significance, which apparently some people do. I also suggest a low-tech way to achieve the same effect.

Randonauts and Coincidences, Fun and Nonsense

People generally seek significance in even the most random of events. This is certainly the case with a new app called Randonautica. It is especially popular among young people who are often called Randonauts. The app asks them to express an idiosyncratic interest and an intention to discover more about it. It then spits out quantum-generated random numbers that are converted to the GPS coordinates of a nearby location, and the Randonauts are then encouraged to locate the spot and see what it is that they can turn up there that is connected to their interest. The intention might be to find out more about a dead relative, some insight into a drawing they've started, the answer to a relationship problem, or whatever.

There are many cases of seemingly uncanny discoveries. A Randonaut interested in death comes across a dead body at his location, someone interested in travel finds an abandoned suitcase packed with clothes, or a woman with many pets finds a sickly cat whining at her GPS coordinates.

The appeal of Randonautica for many is what appears to be the connections between their intentions and the locations to which the quantum generated numbers direct them. Is it evidence that human interests and intentions can affect the quantum-generated random numbers?

The answer is no. This is, I think, just a fascination with the well-known counterintuitive properties and mysteries of quantum mechanics. It's fun to imagine that your intentions have an effect on the numbers generated and that you are somehow communing with the universe and the universe is responding to them by spitting out the locations that it does. The website points to research done on mind–matter interactions by scientists at Princeton University, called the Global Consciousness Project, which started in the late 1990s. Its aim was to determine if psychokinesis exists, if humans' consciousness could influence events, and whether everyday randomness could be disrupted based on a "global consciousness."

No evidence that humans can control or even affect random number generators was found, none at all. No dice (or, rather, all dice).

On the other hand, there is a lot to suggest there are countless ways that aspects of the location (or any location) can be linked to any intention. There are so many names, numbers, acronyms, events, and individuals each of us knows that, if we're so inclined or obsessed and our minds are sufficiently allusive, we can use to weave a plausible connection into the Randonautica-generated location. As I've often said, the most amazing coincidence of all would be the complete absence of all coincidences. Coincidences are ubiquitous. I remember walking around lower Manhattan on September 11, 2002, the first anniversary of 9-11, and reading that Johnny Unitas, the former star quarterback for the Baltimore Colts, had just died. Arguably the number one quarterback in all of football, Unitas wore jersey number 19. Putting 1 and 19 together yielded 911 backward. Voila—nothing.

I won't bore the reader with other meaningless coincidental occurrences. People cherry-pick the most provocative and memorable examples, but, as noted, they're not hard to concoct. We didn't evolve in such a complex world, so we're more impressed by these stories featuring seemingly weird connections than we should be. It might bear reminding that though the probability of almost all of these specific incidents is generally tiny, the probability of some seemingly remarkable and nebulously relevant coincidence occurring at the generated location is very high.

That Randonautica induces people to make up these stories accounts for its fun, as would anything that shakes people out of their routine and purports to reveal hidden aspects of ourselves.

Another low-tech way to do this is simply by taking a different route from home to work each day. If, for example, you live 6 blocks north and 6 blocks east of your place of work, then the number of different direct paths you might take is 924, which is the number of ways of choosing the 6 blocks west you need to walk out of the total of 12 blocks you need to walk. (This is an instance of the notion of combinatorial coefficients, which tell you how many ways there are to choose 6 [or K in general] elements out of 12 [or N in general] elements.) Even though your starting and ending points are the same, each of these 924 walks would take you along at least one block you've never walked before and perhaps elicit a new thought about your interests and intentions at the moment.

More specifically, you might have an ordinary computer or cell phone randomly pick 6 integers out of the first 12 and have these integers determine your walk in the following way. If the 6 numbers chosen are 9, 2, 7, 3, 4, and 10, then you walk SWWWSSWSWWSS, the blocks you walk west determined by the 6 randomly chosen integers. On the remaining 6 blocks, you walk south as indicated.

Finally, the appeal of novelty, randomness, and coincidence brings to mind yet another way to create a link between you and the universe. As I've written elsewhere, it was suggested by Dutch physicist Cornelis de Jager, who suggested you choose three or four numbers that are personally significant to you, say, your birthday, Social Security number, digital password, street address, or whatever. Label these numbers X, Y, Z, and W and consider $X^a Y^b Z^c W^d$, where the exponents range over the values 0, 1, 2, 3, 4, 5, 1/2, 1/3, pi, e, or the negatives of these numbers. (X^{-2} would equal $1/X^2$). Since any one of the four exponents can be any of these 19 numbers, the number of possible choices for a, b, c, and d is by the multiplication theorem, 19^4, or 130,321. Thus, there are this huge number of possible values for $X^a Y^b Z^c W^d$.

Among all these values, there will likely be several that equal or are close to universal constants, such as the speed of light, the gravitational constant, Planck's constant, the fine-structure constant, and so on. Or they might equal important dates in history, the latitude or longitude of your house, or the exact cost of your new car. You might learn, for example, that for your particular choice of X, Y, Z, and W, the number $X^2Y^{1/3}Z^{-3}W^{-1}$ is equal to the sun's distance from the earth. De Jager picked only three numbers and found that the square of his bike's pedal diameter multiplied by the square root of the product of the diameters of his bell and light was equal to 1,816, the ratio of the mass of a proton to that of an electron. Connections to more parochial numbers are also likely. In any case, you might conclude (or hopefully might not) that this shows you to be in inexplicable communion with the universe à la the nonsensical but addictive TV series *Manifest*.

The likelihood of mistaken eyewitnesses is frighteningly high. Lineups are especially likely to result in confidently stated but mistakenly offered identifications. Unfortunately, the Innocence Project is a booming operation.

CALCULATING THE PROBABILITY OF PICKING THE WRONG SUSPECT

"Yes, he's definitely the one I saw that night. I'll never forget that sneer."

Confident but mistaken eyewitness reports during criminal trials can send an innocent man or woman to prison. And new experiments (as well as common sense) indicate that such faulty identifications of suspects are not uncommon.

Before I get to developments on this topic, consider a coin puzzle whose solution is relevant to the issue.

Assume that you have three suspect pennies lined up before you. You're told that one of these pennies, the culprit, lands heads 75% of the time and that the other two, the innocent suspects, are fair coins. You know nothing else about the pennies, but you did previously observe that one of the coins was flipped three times and landed heads all three times. Having witnessed this and realizing that the biased penny is much more likely to behave in this way, you identify this coin as the culprit. How likely are you to be right?

If you were randomly to pick a penny from the lineup, the probability that it would be the culprit would be 1/3, or about 33%. But given that you have this (less than conclusive) information about one of the pennies, what is the probability that it is the culprit?

The answer to the problem, obtained using what is known as Bayes' theorem, is 63%. We revise our probability estimate of that penny's being the culprit upward from 33% to 63% because it's been flipped three times and has landed heads all three times.

The calculations are formally analogous to what we do when we change our estimate of the probability of a suspect's guilt after the testimony of an eyewitness. Identifying a biased coin on the basis of the evidence of three consecutive heads is mathematically the same as identifying a human culprit on the basis of an eyewitness's memory.

There are, of course, many complicating issues in the case of eyewitnesses and suspect lineups. An article by Atul Gawande in *The New Yorker* details the work of Gary Wells, a psychologist at Iowa State University, and others who have noted the alarming error rate among eyewitnesses to crimes. They have discovered a number of factors that significantly influence the likelihood that witnesses will correctly pick the culprit out of a lineup.

Is the lineup of suspects simultaneous or sequential? (Sequential presentation is usually much better.) Do the authorities make an effort to have the others in the lineup match the culprit's description? (More reliable identifications are generally obtained this way.) Is the eyewitness told that the culprit may or may not be in the lineup? (If he's not, he's more likely to pick the person who most resembles the culprit.) Professor Wells has also devised experiments to determine how often eyewitnesses pick the wrong man out of a lineup, and the results are terrifying.

Despite the fact that eyewitnesses are usually quite certain of whom and what they've seen, the probability of a correct identification (after people have seen videotape of a simulated crime, for example) is frequently as low as 60%, and, what's worse, innocents in the lineup are picked up to 20% or more of the time, percentages not much better than in the penny example.

This is not a trivial problem since it's estimated that almost 80,000 people annually become criminal defendants after being picked out of a lineup by eyewitnesses.

Using Bayes' theorem, Wells points out that the base rate likelihood—the initial probability that a suspect is the culprit—greatly affects the subsequent likelihood—the probability that he is the culprit given that he has been picked out of a lineup. If the base rate likelihood is small, eyewitness's identifications need to be almost certain to ensure that the subsequent probability is reasonably high. In the penny problem discussed above, the base rate likelihood of picking the culprit is 33%, the subsequent likelihood 63%. Another useful notion that Wells explores is the information gain from eyewitness identification. This is the difference between the subsequent likelihood that he is the culprit given he's been identified as such and the base rate likelihood that he is.

Again considering the biased penny to be a culprit of sorts, we conclude that the information gain from seeing a penny flipped three times and landing heads all three times is the difference between the probability that the penny is the culprit given that it's landed heads three times (63%) and the initial probability that the penny is the culprit (33%). The information gain is thus 30%.

There are, of course, many other nuances and variations, but the bottom line is that the eyewitness testimony on which people are convicted is sometimes not worth three cents.

Solution to the penny problem: First, let's determine how often we will see three consecutive heads if one of the three pennies is chosen at random and flipped three times. One-third of the time, the culprit coin will be chosen, and, when it is, heads will come up three times in a row with probability 27/64 (3/4 × 3/4 × 3/4), and so 14.1% of the time (.141 = 1/3 × 27/64), the culprit will be chosen and will land heads three times in a row. Two-thirds of the time, a fair coin will be chosen, and, when it is, heads will come up three times in a row with probability 1/8 (1/2 × 1/2 × 1/2), and so 8.3% of the time (.083 = 2/3 × 1/8), a fair coin will be chosen and will land heads three times in a row. The coin selected will thus land heads three times in a row 22.4% of the time (14.1% + 8.3%). Of the 22.4% of the instances where this happens, most occur when the coin is the culprit; specifically [14.1%/ (14.1% + 8.3%)], or 63%, of them do. That is, we will be right 63% of the time if we identify a coin that's landed three times in a row as the culprit among three pennies.

Of course, the penny may cop a plea by pleading insanity and admitting to being unbalanced.

COVID-19 has perhaps had a small positive side effect. It may have heightened awareness of probability and related notions. Below are a standard computation and some related statistical points about base rates, pooling, and sample bias.

COVID-19: False Positives, Fatality Rates, and Base Rates

Fatality rates, inflection points, exponential growth, and false positives are just a few of the mathematical terms highlighted by the COVID-19 curve, germane as well to a large variety of other diseases and conditions. The terms are often misinterpreted, but "false positives," which seems like a clear notion in the daily reporting about testing, has a consequence that is quite counterintuitive. Although it's a staple of popular writing on probability, most people are not aware of just how prevalent false positives can be when the condition being tested is relatively rare, and this has important implications for health policy.

For illustrative purposes, let me describe a standard probabilistic example of this phenomenon. Assume there is a test for condition X (whether a particular cancer, a COVID-19 infection, an intent to commit a criminal or terrorist act, or whatever) that is accurate in the following sense: If someone has condition X, the test will be positive 95% of the time. Furthermore, if one doesn't have it, the test will be negative 98% of the time—in other words, a very accurate and reliable test. Assume further that 0.5%—one out of 200 people—actually have condition X.

Now imagine that you go to your local hospital to be tested and are informed that you have tested positive for X. Many people, including doctors, would assume that your chances of actually having X are very high, around 95% in fact.

The right answer, as noted, will be quite surprising to many. Let's forgo equations, notions like sensitivity and specificity, and related notions and formulas and simply do a little arithmetic. To this end, imagine that in a certain locality, 10,000 tests for X are administered. Of these, how many are likely to be positive? Since we're assuming .5% (or one out of 200 people) will have the condition in question, we conclude that, on average, 50 of these 10,000 people, 0.5% of 10,000, will really have X. And so, since we're assuming that 95% of these 50 with condition X will test positive, we will have, on average, approximately 47.5 positive tests.

But what of the other approximately 9,952.5 (10,000 − 47.5) people who don't have X? Once again, we plug into our assumptions and conclude that 98% of them will test negative, and so about 2% will test positive. This gives us an average total of another 199 positive tests (since 2% of 9,952.5 is about 199).

Thus, of the average total of 246.5 positive tests (199 + 47.5 = 246.5), only 47.5 are true positives, and the rest of them (199) are false positives, and so the probability that you have X when you have tested positive for it is 47.5/246.5, or only about 19%. And remember that this is for a test that was assumed to be quite accurate and reliable.

It's important to state clearly what these two probabilities indicate. If you have X, the test will be positive 95% of the time, but if you test positive for X, you will actually have it only 19% of the time. Let me be the pedantic professor and reiterate the point this way. Imagine the condition X is a deadly cancer. Even if it's the case that those who have X will test positive with probability 95%, the relevant probability for them is the converse probability, the probability that you have cancer given that you tested positive, and this is "only" 19%.

Happily, most of the various tests for COVID are somewhat more accurate than this illustrative example. The percentages will differ for various other conditions X, of course, but the point still holds. When one has a reasonably accurate test and one tests for a relatively rare condition, many (and sometimes most) of the positive tests will be false-positive tests. The relevance of this to COVID-19 tests and especially to antibody tests should be clear. A good number of people who test positive will not really have COVID-19. That is, the tests will be false positives, which would give people, of course, an unwarranted feeling of anxiety. False negatives are also not that uncommon, which would give people, of course, an unwarranted feeling of invulnerability.

As noted, the illustrative numbers above (exactly 95% and exactly 98%) are, of course, more precise than the ones with which epidemiologists generally have to work. I mentioned the fatality rate above, which is simply the number of people who have died from a disease (D) divided by the total number of people who were infected with it (I), or D/I. For COVID-19, this is the fuzziest of fuzzy fractions. Even the numerator, D, is hard to pin down despite the obvious fact that dead is dead. Was COVID a significant factor in the death? And what about the large number of people who died at home? And certainly, the number infected, I, is even more nebulous. What probably large fraction of infected people, many of them asymptomatic, isn't counted? This is the mystery of the denominator. (More generally, denominators are often mysterious or, worse, simply ignored by the media in all sorts of situations since they tend to put matters into perspective.)

None of the above prevents the death toll from being announced daily with authoritative but absurd precision, seemingly correct to the last individual.

Other Issues

Not unrelated is the so-called base rate fallacy when testing for rare conditions. Are there, for example, more surgeons or farmers in the United States who write poetry? Many people will say that there are more surgeons who do so than farmers. What is probably true is that, though a surgeon may be more likely to write poetry than a farmer is, since there are about 100 times as many farmers as surgeons, there are likely to be many more farmer poets than there are surgeon poets.

Or consider a region of the country almost all of whose residents were fully vaccinated against COVID. Seemingly paradoxical, some headlines correctly pointed out that almost half of those hospitalized had been fully vaccinated. Reacting to the headlines, many "anti-vaxxers" claimed that the vaccine was therefore useless. Some of them even sent me "haha-you-are-wrong" emails, but the situation above is as easy to explain as is this one: All the students in a certain region took a difficult exam. About half of those who failed the exam studied hard for it. This doesn't mean studying hard for it was useless, merely that a very large majority of those taking the exam studied hard for it (and thus had a better chance of passing it).

In addition to the prevalence of false positives, false negatives, and vague numbers, there is another problem with any COVID-19 test, whether we're attempting to determine the percentage of infected people or perhaps the percentage of people with antibodies. The figures obtained (at least the early ones) provided an almost textbook case of sample bias. You wouldn't estimate the percentage of alcoholics by focusing your research on bar patrons, estimate the number of sports fans by attending sporting events, or gauge support for stricter gun laws at a shooting range. Who gets tested, especially at first, but people who think they are sick with the virus or were sick with it and have antibodies? The testing of a large random sample of people would have helped immeasurably, but coming up with a large random sample and then inducing them to be tested is daunting. It's hard enough getting people to answer political pollsters on their phones.

One partial solution to the early shortages and slowly forthcoming results of the tests (of any sort) would be, as mentioned, pool testing, whereby the blood or other fluid from a number of people, say, N, is pooled and the pool is then tested. If it tests negative, only one test instead of N is required—a great savings. But if at least one of the N people is sick with COVID (or whatever), the pool will test positive, and each of the N people would then be tested individually for a total of N + 1 tests. The lower the incidence of the condition tested for, the greater the savings on the number of necessary tests.

The best move, as has been endlessly reiterated for a year and a half as of this writing (note: *before* the vaccines), is to wear masks, test, wear masks, test, wear masks, test and trace, preferably before the virus gains a foothold. President Trump, however, questioned the importance of masks and the value of testing. He even argued that tests somehow increase the number of COVID cases. He made this argument, although I doubt he is familiar with 18th-century philosopher Bishop Berkeley, who famously asked, if no COVID-19 tests are given, will there still be COVID-19 cases?

Unfortunately, the same techniques—masks, tests, and tracing—that are effective early on are much less so later. A handheld fire extinguisher can easily put out a stove fire but will be totally ineffective if the house is on fire. South Korea, for example, used their fire extinguishers early, while the Trump administration waited until the house was on fire and even then used faulty fire extinguishers.

Finally, I should note that exponential growth in daily infections is just as exponential when there are 100 cases as when there are 10,000 cases. It's just that in the former situation the graph is not yet as steep and most people have no real understanding of what exponential growth means. As a result of this as well as extreme negligence and cruel politics generally, the Trump administration refused to use the available fire extinguishers until the pandemic had been raging unabated for months.

(Of course, the advent of vaccines in early 2021 changed everything. The obvious best move now is to be vaccinated if you haven't yet been. We don't run into too many people sick with polio or smallpox anymore. I need to note that Trump was also the primary reason for the politicization of COVID-19 vaccinations and the consequent resistance to them. As of this writing, almost three-quarters of a million Americans have died from COVID, many of them the result of negligent homicide.)

Note that the argument below about pre-perpetrators of crime is the same as the argument above about false-positive tests for a disease. This isn't surprising since terrorism can, in this sense, be reasonably conceived of as a disease.

The movie the column below was based on, *Minority Report*, takes place in 2054, but we're essentially there now, well beyond the surveillance measures described in the movie.

FUTURE WORLD: PRIVACY, TERRORISTS, AND PRE-PERPETRATORS

Remember the trial in *Alice in Wonderland* where the sentence precedes the verdict? Not only did last summer's movie *Minority Report* borrow the theme, but so too does the federal government as it hunts for would-be terrorists.

Minority Report takes place in 2054. Its star, Tom Cruise, heads a police unit using futuristic technology and seemingly infallible psychics (or "pre-cogs") to locate people intending to commit murder. The unit's job is to spot these "pre-perpetrators" and then move in, arrest, and incarcerate them before they commit the crime.

The movie, like *Alice in Wonderland*, is a fantasy. Disturbingly reminiscent of the film, however, is a real-life practice that the city of Wilmington, Delaware, adopted this past August. The Wilmington policy calls for police to take pictures of people on the streets near the sites of drug busts. Authorities collect the photos along with the names and addresses of these potential pre-perpetrators and add them to a special law enforcement database. Although most of these bystanders are guilty of no crime, the police deem them more likely than other citizens to commit one in the future. Much more ominous and ambitious than this local initiative in Delaware is the Pentagon's recently proposed techno-surveillance system: Total Information Awareness.

Headed by retired admiral John Poindexter of Iran-Contra notoriety, the initial $10 million for this project (some think the cost will be $240 million over three years) will help set up a system to "detect, classify, ID, track, understand, pre-empt." The objects of these verbs are possible terrorists whom Poindexter hopes to spot before they do any harm. Using supercomputers and data-mining techniques, Total Information Awareness will keep records on credit card purchases, plane flights, emails, websites, housing, and a variety of other bits of information in the hope of detecting suspicious patterns of activity: buying chemicals, renting crop-dusting planes, subscribing to radical newsletters, and so on.

Once again, the aim is to stop pre-perpetrators before they commit any crime—certainly a most worthy goal. The problem is that since the government will collect, integrate, and evaluate extensive personal data on all of us, the system will severely compromise our privacy. On top of this, it's doubtful that it will work anyway.

One objection to it that I want to discuss here stems from probability and the obvious fact that the vast majority of people are not terrorists, murderers, or drug dealers.

A mathematically flavored science fiction scenario about the identification of future terrorists helps make the point. Assume for the sake of the argument that eventually (maybe by 2054) some system of total information gathering becomes so uncannily accurate that when it examines a future terrorist, 99% of the time it will correctly identify him or her as a pre-perpetrator. Furthermore, when this system examines somebody who is harmless, 99% of the time the system will correctly identify him or her as harmless. (This latter probability needn't be the same.)

Now let's say that law enforcement apprehends a person by using this technology. Given these assumptions, you might guess that the person would be almost certain to commit a terrorist act. Right?

Well, no. Even with the system's amazing data-mining powers, there is only a tiny probability the apprehended person will go on to become an active terrorist.

To see why this is so and to make the calculations easy, let's postulate a population of 300 million people, of whom 1,000 are future terrorists. The system will correctly identify, we're assuming, 99% of these 1,000 people as future terrorists. Thus, since 99% of 1,000 is 990, the system will apprehend 990 future terrorists. Great! They'll be locked up somewhere.

But wait. There are, by assumption, 299,999,000 nonterrorists in our population, and the system will be right about 99% of them as well. Another way of saying this is that the system will be wrong about 1% of these people. Since 1% of 299,999,000 equals 2,999,990, the system will swoop down on these 2,999,990 innocent people as well as on the 990 guilty ones, incarcerating them all.

That is, the system would arrest almost 3 million innocent people, about 3,000 times the number of guilty ones. And that occurs, remember, only because we're assuming the system has these amazing powers of discernment! If its powers are anything like our present miserable predictive capacities, an even greater percentage of those arrested will be innocent.

Of course, this is a fiction, and the numbers, percentages, and assumptions are open to very serious questions. Nevertheless, the fact remains that since almost all people are innocent, the overwhelming majority of the people rounded up using any set of reasonable criteria will be innocent. And even though the system proposes only increased scrutiny rather than incarceration for suspected pre-perpetrators, such scrutiny might very well lead throughout time to a detailed government dossier on each of us with little if any increase in security.

There are many ways to combat terrorism without entering a futuristic Wonderland devoid of privacy rights.

Is there something about those women who give birth to a boy in summer that predisposes them to produce more boys? This is a seemingly incomprehensible association, so what stories might we concoct to explain it? This tendency to make up stories is a natural consequence of incomprehensible facts. It would be better if we simply adopted an agnostic attitude toward possible explanations until one is clear, as it is here about summer and boy babies.

SUMMER AND BOYS: A NEW, VERY COUNTERINTUITIVE BIRTHDAY PARADOX

The first is very straightforward. Assume you know that a woman has two children, the older of whom is a boy. Given this knowledge, what is the probability that she has two boys? Let's count the possibilities. The only two are an older boy and younger girl (B-G) or an older boy and a younger boy (B-B). Since these are equally likely, the probability that the woman has two boys is 1/2.

No problem there, but now consider this second scenario. Assume that you know that a woman has two children, at least one of whom is a boy. You know nothing about this boy except his sex. Given this knowledge, what is the probability that she has two boys? You might jump to the conclusion that the answer is again 1/2, reasoning that the sex of one child has no bearing on the sex of the other. This conclusion is incorrect, however, since you don't know whether the boy you know about is the older or the younger child.

Listing two children in the order in which they might be born, we note four possibilities: B-B, B-G, G-B, G-G. We can eliminate G-G since we know that at least one of the two children is a boy. Of the remaining three equally likely possibilities B-B, B-G, or G-B, only one results in two boys. Therefore, the correct conclusion in this case is that the probability that the woman has two boys is 1/3, not 1/2.

This much has long been understood, but the aforementioned paradox is considerably less intuitive.

But What about Summer Births?

Now for the odd result. Suppose that when children are born in a certain large city, the season of their birth, whether spring, summer, fall, or winter, is noted prominently on their birth certificate. The question is, Assume you know that a lifetime resident of the city has two children, at least one of whom is a boy born in summer. Given this knowledge, what is the probability that she has two boys?

On the surface, this appears to be essentially the same as the second scenario—and so it seems that the probability that the woman has two boys, given that at least one of them is a boy born in summer, should remain at 1/3. After all, the season in which a child is born does not seem relevant and should not affect a sibling's sex. Yet if demographers were to collect data on women in this city and focus on those with two children, at least one of whom is a boy born in summer, they would find that 7/15 of these women have two boys.

Why should this be? Is there something about those women who give birth to a boy in summer that predisposes them to produce more boys? Is there some previously unknown genetic/climatological link? All sorts of bizarre theories might be constructed to account for this large increase in probability from 1/3 to 7/15. All of the theories that might be proposed, however, are bound to be wrong. Strange though it may seem, the answer of 7/15 is what we should expect on probabilistic grounds alone.

The Explanation

Let's count the possibilities. Since we're assuming that we know a woman with two children, at least one of whom is a boy born in summer, let's list all the possibilities where this condition is met and see how many of them result in the women having two boys. Abbreviating the four seasons as sp, su, f, and w and listing the older child first, the 15 equally likely possibilities are Bsu-Bsp, Bsu-Bsu, Bsu-Bf, Bsu-Bw, Bsu-Gsp, Bsu-Gsu, Bsu-Gf, Bsu-Gw, Bsp-Bsu, Bf-Bsu, Bw-Bsu, Gsp-Bsu, Gsu-Bsu, Gf-Bsu, Gw-Bsu. (Note we don't count Bsu-Bsu twice.)

Of these 15 possibilities, 7 result in two boys: Bsu-Bsp, Bsu-Bsu, Bsu-Bf, Bsu-Bw, Bsp-Bsu, Bf-Bsu, Bw-Bsu. Thus, knowledge of the summer birth increases our probability estimate from 1/3 to 7/15. If at least one of a woman's two children is a boy, the probability that she has two boys is 1/3, but if at least one of a woman's two children is a boy born in summer, the probability that she has two boys is 7/15.

I reiterate that this calculation shows that the various theories that might be concocted to explain the change in probability from 1/3 to 7/15 are unnecessary and thus are bound to be bogus. As with many random phenomena (in this case, boys and girls being born in equal numbers and more or less uniformly throughout the year), no explanation other than chance is required.

Broader Conclusion

The puzzle illustrates a deeper truth. Even in clear-cut situations, answers, analyses, and explanations may differ depending on subtle differences in phrasings and assumptions. This is all the more true in political and economic situations, which are nowhere near as clear-cut but are even more sensitive to phrasings and assumptions. Coming up with radically different explanations in these more nebulous contexts is not all that surprising, especially when the various actors don't even attempt to approach the issues carefully and in good faith. The likelihood of consensus in such situations is very low, especially in an election season.

CHAPTER THREE

Lies and Logic

"Either it brings tears to their eyes, or else—"

"Or else what?" said Alice, for the Knight had made a sudden pause.

"Or else it doesn't, you know."

—LEWIS CARROLL

ONE DOESN'T HAVE TO BE TOO SNARKY TO REALIZE THAT POLITICS IS A good place in which to state various puzzles about lying. I've always found it interesting that there is this wormhole between the most general and abstract part of mathematics and the often petty and provincial issues of politics. Probably the most well-known logical paradox is the liar paradox, which comes in various forms. "This sentence is false" is one of them. So is "Cretans always lie" when uttered by a lying Cretan. Yet another of many variants involves a crocodile that has captured someone's child and says to the parent, "I will return her to you if you guess correctly whether I will do so or not," and the parent replies, "You won't return my child."

Two more sophisticated examples: Russell's paradox and Quine's paradox. The former asks us to consider the set S, the set of all sets that are not members of themselves. If S is not a member of itself, then by its definition, it is a member of itself, and if it is a member of itself, then by its definition, it is not a member of itself. (Russell resolved this paradox by restricting the notion of a set to a well-defined collection of already existing sets.) Quine's paradox is "yields falsehood when preceded by its quotation" yields falsehood when preceded by its quotation. If you find a quiet place to slowly decode it, you'll realize that this statement also is true if and only if it's false.

A hint at the connection of these self-refuting statements and other logical esoterica to politics is provided in the columns below. One such connection is to the increasingly widespread (wokespread?) allegations of hypocrisy, which can sometimes be partially explained by certain results in mathematical logic. We may wonder, for example, if there is a test or an algorithm that will *always* determine whether a set of statements is consistent and hence whether it harbors any hypocrisy. If the statements are expressed in the formal language of standard (predicate) logic, the answer is absolutely not, so in this sense, someone might be hypocritical and not be aware of it.

Even if the language is the familiar one of propositional logic (truth tables and so on), there are limits to what we can determine. Is there, for example, a test or algorithm that will tell us whether, for a given combination of many simple propositions P, Q, R, . . . connected only by *and*, *or*, and *not*, there is an easy way to assign truth or falsity to the simple propositions in such a way that their combination is true. The answer is that there is no easy way to do so (where "easy" is well defined).

The bottom line is that we probably all subscribe to inconsistent collections of statements and hence are at times hypocritical—sometimes knowingly so, sometimes not. This is all the more the case with natural languages. Moreover, logical terms, rules, and axioms might be precisely stated and clear-cut, but their interpretations needn't be. Indeed, riddles and jokes often have a structure of the form "What has properties A1, A2, and A3?" "It's M, of course." "No, it's N, haha." In mathematical terms, the punch line is the nonstandard model N.

The sections in this chapter and throughout the book focus largely on lies, misconceptions, misunderstandings, puzzles, and paradoxes. Before continuing with these discussions, however, I want to relate a short story I once read that illustrates a misgiving I have about this book. The story involves a man and a woman who are sitting across from one another on a train. The man asks the woman if she is interested in magic tricks, and the woman answers that indeed she is not. The oblivious man continues, "Well, let me explain to you a couple of good ones. Here are two coins. Take one in each hand and I will . . ." I truly hope that I do not come across here as the man on the train. Of course, if I do, you're free to get off the train.

These puzzles illustrate the unhappy, uncivil, and unbreakable partnership between lies and logic. Note the relevance to the (Fib)onacci numbers.

LYING BRAIN TEASERS: POLITICIANS, LIARS, AND MATHEMATICAL PUZZLES

Many recent best-selling books have titles stating directly or indirectly that politicians and political partisans in general are flat-out liars; they fabricate, spin, deceive, and prevaricate.

On the left, these books include Al Franken's *Lies and the Lying Liars Who Tell Them*, Joe Conason's *Big Lies: The Right-Wing Propaganda Machine and How It Distorts the Truth*, and Eric Alterman's *What Liberal Media? The Truth about Bias and the News*. The latter do battle with Ann Coulter's *Slander: Liberal Lies about the American Right*, Dick Morris's *Off with Their Heads: Traitors, Crooks and Obstructionists in American Politics, Media and Business*, and Bernard Goldberg's *Bias: A CBS Insider Exposes How the Media Distort the News* on the right.

These book scuffles bring to mind three tricky puzzles having to do with lies and lying, which, although not very realistic, lead to some important ideas in logic, probability, and number theory.

The Three Puzzles

1. The first sort of puzzle was made popular by the logician Raymond Smullyan, and it concerns, if I may adapt it slightly for my purposes here, a very unusual state in which each of whose politicians either always tells the truth or always lies. One of these politicians is standing at a fork in the road (or a hinge point in politics), and you wish to know which of the two roads leads to the state capital. The politician's

public relations person will allow him to answer only one question, however. Not knowing which of the two types of politician he is, you try to phrase your question carefully so as to determine the correct road to take. What question should you ask him?

2. The politicians in a different state are a bit more nuanced. Each of them tells the truth 1/4 of the time, lying at random 3/4 of the time. Alice, one of these very dishonest politicians, makes a statement. The probability that it is true is, by assumption, 1/4. Then Bob, another very dishonest politician, backs her up, saying Alice's statement is true. Given that Bob supports it, what is the probability that Alice's statement is true now?

3. In a third unusual state, there is another type of politician. This type lies at times but then becomes conscience stricken and makes it a point never to tell two lies in succession. Note that there are 2 possible single statements such a politician can make. They may be denoted simply as T and F, "T" standing for a true statement and "F" for a false one. There are 3 possible sequences of 2 statements, no 2 consecutive ones of which are false—TT, TF, and FT—and there are 5 sequences in which to make 3 such statements—TTT, FTT, TFT, TTF, and FTF. How many different sequences of 10 statements, no 2 consecutive ones of which are false, may a politician in this state utter?

The following are more than hints but less than full explanations. For full understanding, time and a quiet corner may be necessary.

1. You could ask him, "Is it the case that you are a truth teller if and only if the left road leads to the capital?" Another question that would work is, "If I were to ask you if the left road leads to the capital, would you say yes?" The virtue of these questions is that both truth tellers and liars give the same true answer to them albeit for different reasons.

2. First, we ask how probable it is that Alice utters a true statement and that Bob makes a true statement of support. Since both tell the truth 1/4 of the time, these events will both turn out to be true 1/16 of the time (1/4 × 1/4). Now we ask how probable it is that Bob will make

a statement of support. Since Bob will utter his support either when both he and Alice tell the truth or when both lie, the probability of this is 10/16 (1/4 × 1/4 + 3/4 × 3/4). Thus, the probability that Alice is telling the truth given that Bob supports her is 1/10 (the ratio of 1/16 to 10/16).

The moral: Confirmation of a very dishonest person's unreliable statement by another very dishonest person makes the statement even less reliable.

3. Consider all possible sequences of 10 statements, no 2 consecutive ones of which are false. Some of these sequences end with a T and some with an F. There are exactly as many 10-element sequences ending in T as there are 9-element sequences of Ts and Fs since any sequence of either of these two types (9- and 10-element sequences) can be turned into the other type by adding or subtracting a T at the end. Furthermore, there are as many 10-element sequences ending in F as there are 8-element sequences of Ts and Fs since any sequence of either of these two types (8- and 10-element sequences) can be turned into the other type by adding or subtracting a TF at the end. Together, these conditions define the Fibonacci sequence.

Putting these two facts together shows us that there are just as many 10-element sequences of Ts and Fs as there are 8-element sequences and 9-element sequences put together. In particular, there are as many 3-element sequences as there are 2-element and 1-element sequences combined, as many 4-element sequences as there are 3-element and 2-element sequences combined, as many 5-element sequences as there are 4-element and 3-element sequences combined, and so on. But this is the definition of the famous Fibonacci sequence, each of whose terms after the second is the sum of its two predecessors. So the answer to the question is that there are 144 possible 10-element sequences of Ts and Fs having no 2 consecutive Fs. Note that the "Fib" in "Fibonacci" acquires a new resonance.

Maybe to the polemical books I should add *Lying Politicians and the Convoluted Mathematical Truths They Reveal.*

Finally, I should mention that logic enables evasiveness, as well as lying. Asked, "Will you support the expenditure bill or not," a senator might truthfully answer, "Yes."

People who lie all the time aren't the problem. It's those who lie sometimes—that is, all of us, as Smullyan's clever puzzle makes clear. It may be best to read it in a quiet corner.

Logical Liars, Paradoxical Politicians, and Smullyan's Stumper

Lies, lies, lies. Prone to stretching logic, alleging deceit, and passing on gossip, politicians and the media that report on them are, as mentioned, a natural setting for a few classic puzzles involving lying and self-reference.

Superficially political scenarios are also a bit easier to relate to than are the original ones, so I've dressed up some of these conundrums in this more modern garb.

Proceeding from simple to more difficult examples, I'll start with the liar paradox, very well known for millennia. It can result, for example, if a news anchor were simply to announce, "This very statement I'm making is false." If his statement is true, then it's false, and if it's false, then it's true.

Less obvious and more realistic occurrences involving two or more people can also easily arise. If Senator S says that Senator T's comment about the health care bill is false, there is nothing paradoxical about her statement. If Senator T says that Senator S's remark about the issue is true, there is nothing paradoxical about this statement either. But if we combine these two statements, we have a paradox. It's not too hard to imagine a larger collection of such comments from a variety of people, each individually plausible yet leading to an equally potent paradox.

An important principle very similar to that mentioned above is this chestnut: A reporter has two quite knowledgeable sources, A and B. In crucial political situations, A always tells the truth, B always lies, but the reporter has forgotten who is who. The reporter wants to know if Senator S is involved in a certain scandal and for whatever reason can ask only one of his sources, say, by email, a single yes-or-no question. What should it be?

Answer: One solution (there are others) is to ask either source the following question: Are the two statements—(1) you are a truth teller and (2) the senator is involved in this scandal—either both true or both false? The remarkable thing about this question is that both the truth teller and the liar will answer yes if the senator is involved.

If the source is a truth teller, the source will answer yes since both statements are true, and if the source is a liar, the source will answer yes since only one of the two statements is true. A similar argument shows that both sources will answer no if the senator is not involved.

Note that a completely useless question in this situation is, "Are you telling the truth about the senator?" since both liars and truth tellers would answer yes.

The answer above gives rise to an important and general principle. If you want to know if any proposition P is true and your source is a liar or a truth teller, ask him if the two statements—you are a truth teller and proposition P—are either both true or both false. You can trust the answer even if you don't know whether it was given by a truth teller or a liar.

This leads to one of my favorite types of logic problem. The reporter mentioned above might confront an intriguing but more difficult problem, originally formulated in a slightly different scenario by logician Ray Smullyan. Assume said reporter again wants to know whether Senator S is implicated in the scandal, but this time, he has three knowledgeable informants: A, B, and C.

One is a truth teller, one a liar, and one a normal person who sometimes lies and sometimes tells the truth. (They all know each other's status.) The reporter doesn't know who is who, but this time, he can ask two yes-or-no questions, each directed to a single informant, to determine Senator S's involvement. What questions should he ask, and to whom should he direct them?

Answer: Since the previous puzzle showed that we can handle situations with truth tellers and liars, our goal here is to use one of our questions to find an informant who isn't normal. Once we've located him, we've reduced the problem to the previous one.

Thus, the first question should be directed toward A, and it should be, Are the following two statements—you are a truth teller and B is normal—either both true or both false?

Assume that A answers yes. If A is a truth teller or a liar, then we know we can trust the answer, B must be normal, and hence C is not normal. If A is not a truth teller or a liar, then he must be normal, and again we conclude that C is not normal. Either way, a yes answer means C is not normal.

On the other hand, if A answers no and he is a truth teller or a liar, then we can trust his answer and conclude that B is not normal. If A is not a truth teller or a liar, then again we know that B is not normal since A is.

Either way, a no means that B is not normal. If we get a yes, we ask C the second question; if we get a no, we ask B the second question.

And what is the second question? It's the one posed in the first scenario involving only a truth teller and a liar: Are the two statements—(1) you are a truth teller and (2) the senator is involved in this scandal—either both true or both false?

The moral of the story is that complete liars can be as informative as truth tellers. Diogenes, who, as Greek legend had it, spent his life looking for a totally honest man, should have expanded the object of his search. A totally dishonest man would have done just as well. The problem is with those pesky critters who sometimes lie and sometimes tell the truth.

A similar point can be made about "anti-psychics" who are wrong much more often than they're right. Should there be such folks, one should follow them around and simply negate everything they say.

This brief excerpt is from my book *I Think, Therefore I Laugh* published by Columbia University Press. It was inspired by Groucho's old TV show *You Bet Your Life*.

GROUCHO MEETS RUSSELL

Bertrand Russell and Groucho Marx were both in their own way concerned with the notion of self-reference. Furthermore, Russell's theoretical skepticism contrasts with Groucho's streetwise brand, as do Russell's aristocratic anarchist tendencies with Groucho's more visceral anarchist feelings. I try to illustrate these points in the following dialogue between the two. Groucho Marx and Bertrand Russell: What would the great comedian and the famous mathematician-philosopher, both in their own way fascinated by the enigmas of self-reference, say to each other had they met. Assume for the sake of absurdity that they are stuck together on the 13th metalevel of a building deep in the heart of Madhattan.

Groucho: This certainly is an arresting development. How are your sillygisms going to get us out of this predicament, Lord Russell. (Under his breath: Speaking to a Lord up here gives me the shakes. I think I'm in for some higher education.)

Russell: There appears to be some problem with the electrical power. It has happened several times before, and each time everything turned out quite all right. If scientific induction is any guide to the future, we shan't have long to wait.

Groucho: Induction, schminduction, not to mention horsefeathers.

Russell: You have a good point there, Mr. Marx. As David Hume showed 200 years ago, the only warrant for the use of the inductive principle of inference is the inductive principle itself, a clearly circular affair and not really very reassuring.

Groucho: Circular affairs are never reassuring. Did I ever tell you about my brother, sister-in-law, and George Fenniman?

Russell: I don't believe you have, though I suspect you may not be referring to the same sort of circle.

Groucho: You're right, Lordie. I was talking more about a triangle and not a cute triangle either. An obtuse, obscene one.

Russell: Well, Mr. Marx, I know something about the latter as well. There was, you may recall, a considerable brouhaha made about my appointment to a chair at the City College of New York around 1940. They objected to my views on sex and free love.

Groucho: And for that they wanted to give you the chair?

Russell: The authorities, bowing to intense pressure, withdrew their offer, and I did not join the faculty.

Groucho: Well, don't worry about it. I certainly wouldn't want to join any organization that would be willing to have me as a member.

Russell: That's a paradox.

Groucho: Yeah, Goldberg and Rubin, a pair o' docs up in the Bronx.

Russell: I meant my sets paradox.

Groucho: Oh, your sex pair o' docs. Masters and Johnson, no doubt. It's odd a great philosopher like you having problems like that.

Russell: I was alluding to the set M of all sets that do not contain themselves as members. If M is a member of itself, it shouldn't be. If M isn't a member of itself, it should be.

Groucho: Things are hard all over. Enough of this sleazy talk though. (Stops and listens.) Hey, they're tapping a message on the girders. Some sort of a girder code, Bertie.

Russell: (Giggles) Perhaps we should term it a Godel code, in honor of the eminent Austrian logician Kurt Godel.

Groucho: Whatever. Be the first contestant to guess the secret code and win $100.

Russell: I shall try to translate it. (He listens intently to the tapping.) It says, "This message is . . . This message is . . ."

Groucho: Hurry and unlox the Godels, Bertie boy, and st-st-stop with the st-st-stuttering. The whole elevator shaft is beginning to shake. Get me out of this ridiculous column.

Russell: The tapping is causing the girders to resonate. "This message is . . ."

A LOUD EXPLOSION.

THE ELEVATOR OSCILLATES SPASMODICALLY UP AND DOWN.

Russell: ". . . is false. This message is false." The statement as well as this elevator is ungrounded. If the message is true, then by what it says, it must be false. On the other hand, if it's false, then what it says must be true. I'm afraid that the message has violated the logic barrier.

Groucho: Don't be afraid of that. I've been doing it all my life. It makes for some ups and downs and vice versa, but as my brother Harpo never tired of not saying, why a duck?

Suffering perhaps from a logician's occupational obsession or a busman's holiday, I offer this in the spirit of self-referential oddities in daily life. One common example occurs when a person makes a statement and nonverbally negates with a smirk or an eye roll. Less well known is the following example mentioned in my book on the stock market.

TRUE IF AND ONLY IF FALSE IN THE MARKET AND IN LIFE

As noted, the liar paradox and its descendants and variants seems particularly apt today when so much of the news on the internet and in the larger analog world seems so strange, almost paradoxical. A couple of old standards: Groucho Marx famously vowed that he'd never join a club that would be willing to accept him as a member. Epimenides the Cretan exclaimed, "All Cretans are liars." The prosecutor demands, "You must answer yes or no. Will your next word be 'no'?"

I even have my own joke about self-referential paradoxes, but this sentence isn't it.

One more: The author of a popular investment book suggests that we follow the hundreds of thousands of his readers who have gone against the crowd. The latter brings to mind a quantitative analogue of the above paradoxes. It's not itself a paradox, but it does have a distinctly paradoxical whiff.

It is an odd claim about the so-called Efficient Market Hypothesis (EMH). This is the well-known hypothesis about the stock market that maintains that stock prices more or less immediately and efficiently take into account all available information impacting the stock(s). (There are weaker versions of the hypothesis.) If the EMH is true, one of its consequences is that consistently "beating the market" is impossible since stock prices very quickly rise or fall because they very quickly adjust to and reflect all available information. Since the totality of all this information is quasi-random, largely unpredictable, and beyond our complexity horizon, so are stock prices on the whole.

So, is the EMH true or not? Is the market efficient or not? Let's assume investors believe the EMH and ask how they would likely behave, what they would likely do. The answer arguably is that they would do nothing. There would be no reason to quickly buy on good news or quickly sell on bad news since they believe the stock price would have already adjusted to the news whatever it is. Thus, if a majority of investors believed the EMH, they would pass up what seemed like a good buy, and the result would be that the market would not so quickly respond to investors' actions, and it would become inefficient.

On the other hand, let's assume that investors do not believe the EMH and ask how they would likely behave, what they would likely do. The answer arguably is that they would scramble to take advantage of inefficiencies in the market, of stocks that are at least temporarily underpriced or overpriced, and quickly buy or sell them. The result would be that by their efforts, they would make the market efficient. Thus, if a majority of investors did not believe the EMH, they would jump on perceived good buys, and the market would quickly respond to these investors' actions, and it would become efficient.

Putting these two conclusions together, we get what I think is the paradox of the EMH. The EMH is true if and only if most investors believe it to be false. Equivalently, the EMH is false if and only if most investors believe it to be true. Or so says Epimenides, the stock market analyst.

Another Issue

More general is the multi-agent problem, a revealing illustration of which is the following. Ask a group of people to guess a number that will be determined by them. Tell them you will give a large prize, say, $10,000, to whoever of them guesses the number that is 80% of the average of all the numbers guessed by people in the group. Sounds simple. People might figure that most people would guess numbers at random, and thus the average would likely be around 50, and so they might guess a number around 40, which is 80% of 50. But thinking about the matter more critically, some in the group would look around and decide that the others in the group would also guess around 40, and so they might decide to guess a number around 32, which is 80% of 40. Seeing the very intelligent glimmer in the eyes of many of the others in the group, some would decide that the others would also guess around 32, and so instead they might guess around 25.6, which is 80% of 32.

This reevaluation before guessing can be iterated, and if people decided the others' insight into the problem is as penetratingly trenchant as theirs is, they would guess 0, which of course is 80% of 0 and can't be improved on. So 0 is the correct answer and incidentally is the so-called Nash equilibrium point. But here's the rub. Anybody smart enough to get the right numerical answer in this sense would very likely get the wrong social answer since it would be a good bet that most of the others wouldn't be anywhere near as perspicacious. Alas, figuring people is often more difficult than figuring numbers.

It's better to say "X is this good thing Z" rather than "X is not this bad thing Y." A political trick, for example, is to get your opponent to deny that he did this or that reprehensible thing and then haplessly try to refute your manufactured details.

OH NO. DENIALS AND CONDITIONAL STATEMENTS OFTEN COUNTERPRODUCTIVE

Denying a proposition may give it added credibility, especially when the denial appears widely on television and on large social media platforms. A study by University of Michigan psychologist Norbert Schwarz supports this claim. Schwarz copied a flier put out by the Centers for Disease Control (CDC) intended to combat various myths about the flu vaccine. It listed a number of common beliefs about the vaccine and indicated whether they were true or false. He then asked volunteers to read the flier. Some of them were old, some young, but shortly thereafter, he found that many of the older people mistakenly remembered almost a third of the false statements as being true, and after a few days, young and old alike misclassified 40% of the myths as factual.

Even worse was that people now attributed these false beliefs to the CDC itself! In an effort to dispel misconceptions about the vaccine, the CDC had inadvertently lent its prestige to them. In many cases, truth and elucidation can actually strengthen misconceptions and make them more psychologically available or accessible.

Related studies by Kimberlee Weaver of Virginia Polytechnic University and others have shown that being repeatedly exposed to information from a single source, say, a president of the United States, is often tantamount to hearing it from many sources. People simply forget where they heard something after a while, and the repetition makes it more psychologically available and hence credible.

In another simple experiment, groups of students were given an email to read, some groups receiving a version that seemed to have a software bug in it since the same crucial paragraph was reprinted two more times at the bottom of the email. Those students with the repeated information were more persuaded by the email and overestimated the information's general appeal than students who had read the nonrepetitive email. Trump on Twitter is once again an exemplar of this phenomenon.

The difficulty in processing denials is probably part of the reason for their frequent ineffectiveness. Complexity and logical connectives get lost in transmission. (Quick, what does the following sentence mean? "It's not the case that Waldo denied that evidence was lacking that he did in fact fail to do X.") This suggests that people often mentally transmute a denial into an assertion. They hear "X is not this bad thing Y," and soon enough, what remains with them is the link between X and Y, which eventually becomes "X is Y."

A specific lesson from this research seems clear. Denials of assertions should in general not repeat the assertions. It's better to say "X is this good thing Z" rather than "X is not this bad thing Y." Needless to add is that critics of Trump too often repeat his many lies and give them vastly more exposure than they would otherwise get. Moreover, any search for X on the internet will yield a prominent mention of Y as well.

Such confusions apply not only to denials but to any statement that requires a degree of logical processing. For example, what might be remembered of a conditional statement such as "If A, then B," depending on whether A or B strikes an emotional cord, is the converse statement "If B, then A." Demagogues often encourage this sort of sloppy, inattentive thinking. "If you're a card-carrying communist, you're an atheist," morphs into "You, Mr. Nonbeliever, are a communist." And, of course, the use of logical terms like "necessary," "sufficient," "at least," "at most," "all," and "every" are sure to bring about confusion among many.

In general, the inability to handle statements of any complexity has the very unfortunate consequence of reducing much of political discourse to slogans and insults. That describes much of Twitter too, where competing threads of assertions and denials often lead to meaningless brouhahas and the spread of nonsensical rumors.

In any case, I want to here deny that my books and articles have been nominated for the Nobel Prize in Literature. Okay, okay, the denial has to possess some minimal plausibility.

Life reversals, environmental catastrophes, personal disappointments, taxes, and economic downturns—these all illustrate the often incompatible attitudes we have to future risks and rewards. So do a couple of simple paradoxes that have a logical zing.

THE ELLSBERG AND ST. PETERSBURG PARADOXES: RISKS, UTILITY, AND OUR DESIRE FOR CERTAINTY

Money, money, money. Everyone wants more, but, alas, the second million won't do for you what the first one did, nor will you be as willing to take the same risks to get it.

It was Daniel Bernoulli, the 18th century Swiss mathematician, who wrote that people's enjoyment of any increase in wealth (or regret at any decrease) is "inversely proportional to the quantity of goods previously possessed." The more dollars you have, the less you appreciate getting one more and the less you fear losing one.

What's important is the "utility" to you of the dollars you receive, and their utility drops off, often logarithmically, as you receive more of them. Gaining or losing $1 million means much more to most people than it does to Warren Buffett or Bill Gates. People consider not the dollar amount at stake in any investment or game but rather the utility of the dollar amount for them.

Note that the declining average utility of money provides part of the rationale for progressive taxation and higher tax rates on greater wealth.

A less weighty illustration than progressive taxation is provided by a recent British study of the show *Who Wants to Be a Millionaire*. It confirms that contestants behave as considerations of utility would suggest. Once they've reached a high rung on the winnings ladder, they more often quit while ahead rather than risk falling to a much lower level.

On the show and in general, people tend to be risk averse and usually choose the sure thing.

Likewise, despite the unequal expected values of the following alternatives, almost everyone offered the choice between (1) $100 million or (2) a 1% chance at $15 billion will choose the sure $100 million. (This is despite the fact that 1% of $15 billion is $150 million.) And most people would choose a sure $30,000 rather than an 80% chance of being given $40,000, although the expected value of the latter is $32,000. Note, however, that if money and the above choice were offered every day for a year, it would be wiser to always choose an 80% chance of $40,000.

The St. Petersburg Paradox

The notion of utility also resolves the famous St. Petersburg paradox. The paradox usually takes the form of a game requiring that you flip a coin repeatedly until a tail first appears. If a tail appears on the first flip, you win $2. If the first tail appears on the second flip, you win $4. If the first tail appears on the third flip, you win $8, and, in general, if the first tail appears on the Nth flip, you win 2^N dollars. How much would you be willing to pay to play this game?

In a sense, you should be willing to pay any amount to play this game. To see why, recall that the probability of a sequence of independent events such as coin flips is obtained by multiplying the probabilities of each of the events. Thus, the probability of getting the first tail, T, on the first flip is 1/2; of getting a head and then the first tail on the second flip, HT, is $(1/2)^2$, or 1/4; of getting the first tail on the third flip, HHT, is $(1/2)^3$, or 1/8; and so on.

Multiplying the probabilities for the various outcomes of the St. Petersburg game by the size of these outcomes and adding these products gives us the expected value of the game: ($2 × 1/2) + ($4 ×1/4) + ($8 × 1/8) + ($16 × 1/16) + . . . ($2^N × [1/2]^N$) + ... Each of these products

82

equals $1, there are infinitely many of them, and so their sum is infinite, and this is why it can be argued that you should be willing to pay any price to play this game. No matter how much you bet each time you play, you'll still come out way ahead on average.

But the failure of expected value to capture human intuitions becomes clear when you ask yourself why you'd be reluctant to pay even a measly $1,000 for the privilege of playing this game. That measly $1,000 is of more utility to you than are the billions of dollars that are only a remote possibility.

(Similar sorts of declining utility characterize other sorts of wealth. Someone who's publicity rich will get a much smaller kick out of an article by him or her in a magazine, say, than someone who's never been published before. Someone who's traveled extensively all over the world will get less out of a few days in Florence than will someone else who's never left Torrance. And someone who's had sex with many partners will get, well, you get the idea.)

The Ellsberg Paradox

To further discombobulate yourself, consider the so-called Ellsberg paradox, named after Daniel Ellsberg of Pentagon Papers fame. Imagine a large urn—mathematicians like large urns almost as much as they like coins and dice—containing 300 marbles, exactly 100 of them red and the other 200 of them black and yellow in unknown proportions (i.e., from 0 to 200 black, the others yellow).

You're asked to choose option A or option B. Under A, you'll receive $1,000 if you pick a red marble from the urn, whereas under B, you'll receive $1,000 if you pick a black marble. Which option would you take?

You're also asked to choose between option C and option D. Under C, you'll receive $1,000 if you pick a red or a yellow marble from the urn, whereas under D, you'll receive $1,000 if you pick a black or a yellow marble from the urn. Which option would you take here?

You'll prefer option A to option B exactly when you think picking a red marble from the urn is more probable than picking a black marble. Likewise, you'll prefer option C to option D exactly when you think picking a red or a yellow marble from the urn is more probable than picking a black or a yellow marble.

It would seem too that if you prefer option A to option B, you should also prefer C to D. And, if you prefer option D to option C, you should also prefer B to A.

The only problem is that this is not what people actually do. Most prefer option A to option B (presumably because they know for sure that there are 100 red marbles in the urn), but they prefer option D to option C (presumably because they know for sure that there are 200 black and yellow marbles in the urn). People are averse to uncertainty and choose the sure thing even when their behavior is inconsistent and violates the usual axioms of utility theory and subjective probability.

A more common, less subtle example of the sometimes irrational desire for certainty is the preference of many people for precision even when it's unwarranted. Too many seem to believe that much better than a rough but reasonable estimate is a precise but quite dubious figure. The U.S. Census is an unfortunate example. Many, especially Republican lawmakers, would rather have an exact but serious undercount of the population than allow statistical surveys in selected neighborhoods and regions to get a reasonable estimate of the uncounted.

Interestingly, this preference for precision and certainty even when they're unjustified leads to a way I have found to distinguish between those with a math or scientific background and those innocent of such disciplines. If someone in a TV advertisement cites a very precise number, say, 59.61% of customers or a weight of 16.78 pounds, those with a technical background will usually be quite wary of these numbers and the digits after the decimal point, whereas those without such a background are more likely to be impressed by the figure, even if the precision is 99.999% flapdoodle.

Most of us don't like risk, imprecision, and uncertainty. That's too bad because there's no shortage of them.

You think you know something. Maybe you do, and maybe you don't. Despite what philosophers have believed since the ancient Greeks, knowledge is not justified true belief, and that can sometimes have real-world consequences. One such illustration is the 2020 Democratic presidential primary.

JUSTIFIABLY BELIEVING THAT SOMETHING IS TRUE DOESN'T MEAN YOU KNOW IT

I've always liked stories that depended on mistaken identity, a very old theme in general. Having a degree in mathematical logic, I was also drawn to the subject on a more theoretical level, on which lies Gettier's Paradox.

Since Plato and the ancient Greeks, knowledge has been taken by many philosophers of science to be justified true belief. A subject S is said to know a proposition P if P is true, S believes that P is true, and S is justified in believing that P is true. The philosopher Edmund L. Gettier showed in 1963 that these three ancient conditions are not sufficient to ensure knowledge of P. His counterexamples to a straightforward under-standing of knowledge are paradoxical and seem particularly prevalent in politics. For me, this is part of their appeal since politics and mathematical logic occupy such different realms of cognitive space.

To provide a topical one, consider the 2020 election. Joe Biden, Bernie Sanders, and Elizabeth Warren in the early spring before the 2020 nominating convention were certainly evaluating their chances to win the Democratic nomination for the presidency. Biden was unimpressive on the campaign trail at that point, while Sanders and Warren had a lot of energetic progressive support, and so Biden had evidence for the following compound proposition: Proposition (1): Sanders or Warren will get the nomination, and there is a little clock that might help them out mounted in their lecterns during the early debates. Biden's evidence for (1) might be that the polls were showing that both Sanders and Warren were the clear front-runners at the time. He also noticed the clock as he looked to the left and right toward their lecterns during the early debates. If (1) is true, it implies Proposition (2): The person who will get the nomination is standing behind a lectern with a little clock mounted in it. Biden saw that (1) implied (2) and thus accepted (2) on the basis of (1), for which he had good evidence. Clearly, Biden was justified in believing that (2) was true.

So far, so good. But unknown to Biden at that time was that he, not Sanders or Warren, would get the Democratic nomination.

Stay with me. Also unknown to him was that his lectern also had a very small clock mounted in it that he hadn't noticed. Proposition (2), the eventual Democratic nominee had a little clock in the lectern, was thus true even though (1), from which it was inferred, was false. Now all the following are true: (2) was true, Biden believed (2) was true, and Biden was justified in believing (2) was true. But, of course, it is quite clear that Biden didn't really know that (2) was true since (2) was true in virtue of the fact that his lectern had an unnoticed clock mounted in it. Thus, justified true belief does not constitute knowledge.

Speaking of clocks, I should mention Bertrand Russell's stopped clock as a somewhat trivial example. The clock might be stopped at 8:11 and thus indicate that it was 8:11 when you glanced at it as you passed by. If the time happened to be 8:11, your belief that it was 8:11 would be justified and true, but it couldn't very well be said that you knew it was 8:11.

Clocks are, of course, not the issue. More generally, we all believe true statements with good justifications, but in many cases, we can't characterize what we believe as real knowledge. Consider a different but common sort of example. You see a dog that is groomed or disguised to look like a sheep standing out in a hilly field. It might be the case that in the field on the other side of a hill stands a real sheep that you can't see. You would believe there was a sheep in the field, the belief would be justified, and the belief would be true. Still, you could not be said to know there was a sheep in the field. For sheep in disguise, you can substitute a Republican elephant or a Democratic donkey to induce people to believe they know something they don't. (Compare Justice Kavanaugh's claim in his defense that there was a doppelgänger who was Blasey Ford's real assailant.)

In any case, if justified true belief does not constitute knowledge, the question remains, What does? One answer, due to the philosophers Fred Dretske (with whom I took several courses at the University of Wisconsin) and Robert Nozick, involves the notion of subjunctive tracking or counterfactual conditionals. But this is perhaps getting too deep into the philosophical weeds.

Suffice it to say that given all the varieties of fake news (and true news), epistemology and politics should not be such strangers. Abstract and visceral though they respectively may be, they do dance together more often than many realize.

It's often very difficult to distinguish relatively easy mathematical problems from ones that have remained unsolved for centuries.

PROVE THIS, WIN $1,000,000! THE GOLDBACH AND COLLATZ CONJECTURES AND A DOABLE PUZZLE

One generally doesn't speak the words "prime numbers" and "seven-figure prizes" in the same breath. But don't tell that to the publishers of *Uncle Petros and Goldbach's Conjecture*, an engaging first novel by Greek author Apostolos Doxiadis.

Before getting to the money, here's a quick synopsis of the story: The narrator tells of his Uncle Petros, whom he initially thinks of as the eccentric black sheep of the family. Slowly, Uncle Petros is revealed to be a character of complexity and nuance, having devoted his considerable mathematical talents and much of his life to a futile effort to prove a classic unsolved problem. His solitary efforts give one a taste of the delight and the despair of mathematical research.

Goldbach's conjecture, Uncle Petros's holy grail, is startlingly simple to state: Any even number greater than 2 is the sum of two prime numbers.

Remember that a prime number is a positive whole number that is divisible only by two numbers: itself and 1; thus, 5 is a prime, but 6, which is divisible by 2 and 3, is not. The number 1 is not considered prime.

Check out the claim. Pick an even number at random and try to find two primes that add up to it. Certainly, 6 = 3 + 3, 20 = 13 + 7, and 97 + 23 = 120. (This, of course, is not a proof.) The conjecture that this works for every even number greater than 2 was proposed in 1742 by Prussian mathematician Christian Goldbach. To this day, it remains unproven despite the efforts of some of the world's best mathematicians.

For those with a desire to prove something (albeit something easier), try this. (Your reward will be the satisfaction of understanding, which is worth more than money.) Pick any 10 numbers between 1 and 100. There will always be two subsets of these 10 numbers whose sums are equal. Thus, for example, if you were to choose 51, 11, 81, 68, 73, 87, 23, 29, 25, and 94, as I just did using a random number generator, you would soon observe that $25 + 51 + 29 = 94 + 11$. The claim is that this works for every 10 numbers you choose.

Prove it! Likewise, if you were to pick 20 whole numbers between 1 and 50,000, you would always find two subsets of these 20 numbers whose sums were equal.

Number theory, the branch of mathematics that studies prime numbers and other ethereal aspects of the integers (whole numbers), contains many problems that are easy to state yet resistant, so far, to the efforts of all. The twin prime conjecture is another: There are an infinite number of prime pairs, prime numbers that differ by 2. Examples are 5 and 7, 11 and 13, 17 and 19, 29 and 31, and, presumably, infinitely many more.

Of more contemporary origin is the so-called Collatz conjecture, which is sometimes called the $3\times + 1$ problem. Choose any whole number. (Take 13, for example.) If it is odd, multiply it by 3 and add 1. (3 times 13 plus 1 equals 40.) If it is even, divide it by 2. (40 divided by 2 is 20.) Continue this procedure with each resulting number, and the conjecture is that the sequence thus generated always ends up 4, 2, 1, 4, 2, 1, 4, ... The sequence starting with 13 produces 13, 40, 20, 10, 5, 16, 8, 4, 2, 1, 4, 2, 1, ... In fact, every number that's been tried (up to about 27 quadrillion) ultimately cycles back to 4, 2, 1, but there is still no proof that every number does.

Like the recently proved Fermat's Last Theorem (which says that the equation $x^n + y^n = z^n$, where x, y, z, and n are all integers, has no integer solutions for $n > 2$), these conjectures are tantalizing and can sometimes become, if one is not careful, all-consuming obsessions. Such an obsession is the fate of Uncle Petros. His brothers think little of him and of his quixotic attempt to prove Goldbach's conjecture.

Petros in turn has disdain for their petty concern with the family business in Athens. Interestingly, Petros also has a low regard for applied mathematics, which he compares to glorified "grocery bill" calculations and which, he believes, shares none of the austere beauty of pure number theory.

Theory versus Practical

This mutual contempt between mathematicians and more practical sorts has a long history. The British mathematician G. H. Hardy, a colleague of the fictional Uncle Petros in the book, exulted in the uselessness of mathematics, particularly number theory.

Happily, this adversarial attitude has softened in recent years, and even number theory, arguably the most impractical area of math, has found important applications. Cryptographic codes, which enable the transfer of trillions of dollars between banks, businesses, and governments, depend critically on number theory.

They depend, in particular, on the simple fact that multiplying two large prime numbers together is easy, but factoring a large number (say, one having 100 digits) into prime factors is extraordinarily difficult and time consuming.

Finally, I come to the million-dollar contest. The U.S. publisher of *Uncle Petros and Goldbach's conjecture* has promised $1 million to the first person to prove the conjecture, provided the proof appears in a reputable mathematics journal before 2004.

The late, great number theorist Paul Erdos used to offer small monetary prizes to anyone solving this or that problem, but he didn't have to pay up often. If I were the publisher, I wouldn't worry about the offer's financial risk, but I would be apprehensive about the torrent of false proofs that will soon be heading their way.

Solution to the Small Challenge

From the integers between 1 and 100, choose any subset of 10 numbers. Call it S. How many subsets does S have?

A preliminary observation is that a set with 2 numbers in it, say, a and b, has 2^2 subsets: {a}, {b}, {a, b}, and the empty subset {}. A set with 3 numbers in it, say, a, b, and c, has 2^3, or 8, subsets: {a}, {b}, {c}, {a, b}, {a, c}, {b, c}, {a, b, c}, and the empty subset {}. More generally, one can show that a set with n numbers in it has 2^n subsets, $2^n - 1$ of them nonempty.

Since the set S contains 10 numbers, it has $2^{10} - 1$, or 1,023, subsets. But how many possible sums are there for the numbers in each of these 1,023 subsets of S?

Even if for S you chose the 10 largest numbers, 91, 92, 93, . . . 100, the sum is less than 1,000 (955, to be exact). Thus, for any subset of S, the sum would certainly be less than 1,000. That, in turn, means that whatever S you choose, there are fewer than 1,000 possible sums for the numbers in each of its subsets. Since 1,023 is greater than 1,000, there are always more subsets of S than there are possible sums for the numbers in each of the subsets. Thus, at least two of the 10-element subsets of S must have the same sum.

Nestled between lies and logic are con games. What may seem like clairvoyance, for example, may be a consequence of a card trick. A most remarkable one is due to Martin Kruskal, although he is not responsible for the use to which I put it.

A CLEVER CARD TRICK AND A RELIGIOUS HOAX
Martin Kruskal, a renowned mathematician and physicist at Rutgers University, died in December 2006. Of his many accomplishments, there is an intriguing trick that almost anyone can appreciate.

I explain it here, and, prompted by April Fool's Day, I also sketch a sort of biblical hoax based on it that I first proposed in my 1998 book *Once Upon a Number*.

Kruskal's trick can be most easily explained in terms of a well-shuffled deck of cards with all the face cards removed. The deck has 1s (aces) through 10s only. Imagine two players, Hoaxer and Fool. Hoaxer asks Fool to pick a secret number between 1 and 10.

For illustration, let's assume Fool picks 7. Hoaxer goes on to instruct Fool to watch for the card with his secret number—in this case, the 7th card in the deck—as Hoaxer slowly turns over the cards one by one.

When the card with the secret number is reached, Hoaxer directs Fool to take the value of this card as his new secret number and to repeat the process.

Thus, when the 7th card is reached—let's assume it's a 5—Fool's new secret number becomes 5, and so he watches for the next card corresponding to this new secret number. That is, he watches for the 5th card succeeding it in the deck.

Hoaxer continues to slowly turn over the cards one by one. When the 5th succeeding card is reached—let's assume it's a 9—Fool's new secret number becomes 9, and so he watches for the card corresponding to this new secret number. That is, he watches for the 9th card succeeding it in the deck and so on.

As they near the end of the deck, Hoaxer turns over a card and announces, "This is your new secret number," and he is almost always correct.

Why Does It Work?

The deck is not marked or ordered in any way, there are no confederates, there is no sleight of hand, and there is no careful observation of Fool's reactions as he watches the cards being turned over.

How does Hoaxer accomplish this feat? The answer is cute.

At the beginning of the trick, Hoaxer picks his own secret number. He then follows the same instructions he's given to Fool. If he picked a 3 as his secret number, he watches for the 3rd card and notes its value—say, it's a 6—which becomes his new secret number. He then looks for the 6th card after it—say, it's a 4—and that becomes his new secret number and so on.

Even though there is only 1 chance in 10 that Hoaxer's original secret number is the same as Fool's original secret number, it is reasonable to assume—and it can be proved—that sooner or later, their secret numbers will coincide. That is, if two more or less random sequences of secret numbers between 1 and 10 are selected, sooner or later they will, simply by chance, lead to the same card.

Furthermore, from that point on, the secret numbers will be identical since both Fool and Hoaxer are using the same rule to generate new secret numbers from old. Thus, all Hoaxer does is wait until he nears the end of the deck and then turn over the card corresponding to his last secret number, confident that, by that point, it will probably be Fool's secret number as well.

Note that the trick works just as well if there is more than one Fool or even if there is no Hoaxer at all (as long as the cards are turned over one by one by someone).

If there is a large number of people and each picks his or her own initial secret number and generates a new one from the old one in accordance with the procedure above, all of them will eventually have the same secret number, and thereafter the numbers will move in lockstep.

My Proposal for a Religious Hoax

Consider now a holy book with the compelling property that no matter what word from the early part of the book is chosen, the following procedure always leads to the same climactic and especially sacred word.

Begin with whatever word you like; count the letters in it (say, this number is X), proceed forward X words to another word, count the letters in it (say, this number is Y), proceed forward Y words to another word, count the letters in it (say, this number is Z), and keep on doing this until the climactic and especially sacred word (say, "God" or "heaven") is reached. (The letter count of each word plays the role of the numbered cards.)

It's not too hard to imagine frenzied checking of this procedure using word after word from the early part of the holy book and the increasing certainty among some that divine inspiration is the only explanation for the fact that the procedure always ends on the sacred word. If the generating rule were more complicated than the simple one above, the effect would be even more mysterious.

The reader can experiment with his or her own examples or check out the August 1998 issue of *Scientific American*, where puzzle meister Martin Gardner, who graciously blurbed my book, came up with an elegant illustration of its proposal for a religious hoax.

Martin Kruskal, I should note, was innocent of perpetrating any hoaxes. He was simply a very fine applied mathematician.

In any case, Happy April Fool's Day to Hoaxers and Fools alike.

An example of a cheerier use of the Kruskal trick that shows that the Declaration of Independence guarantees "happiness."

DOES THE DECLARATION OF INDEPENDENCE GUARANTEE HAPPINESS?

Pick *any* word from the first two lines of the Declaration of Independence (see the excerpt below). Call it your special word. Count the number of letters in it and move forward that number of words to your next special word. For example, if you picked the word "course," which has six letters, as your special word, then you would move ahead six words to "necessary," which becomes your next special word. It has nine letters, so you would move ahead nine words brings you to your next special word, "which." Continue to follow this rule until you can't go any farther. Explain why no matter what word you pick initially, your last special word will be "happiness." Does the Declaration guarantee happiness?

Here's the example:

When in the **Course** of human events, it becomes **necessary** for one people to dissolve the political bands **which** have connected them with **another**, and to assume among the powers **of** the **earth**, the separate and equal **station** to which the Laws of Nature **and** of Nature's **God** entitle them, **a decent** respect to the opinions of **mankind** requires that they should declare the **causes** which impel them to the **separation**. We hold these truths to be self-evident, that all **men** are created **equal**, that they are endowed **by** their **Creator** with certain unalienable Rights, that among **these** are Life, Liberty and **the** pursuit of **Happiness**."

Now you try it with an initial word different from "course" and check that you still end up with "Happiness." Here's the passage again.

"When in the Course of human events, it becomes necessary for one people to dissolve the political bands which have connected them with another, and to assume among the powers of the earth, the separate and equal station to which the Laws of Nature and of Nature's God entitle them, a decent respect to the opinions of mankind requires that they should declare the causes which impel them to the separation. We hold these truths to be self-evident, that all men are created equal, that they are endowed by their Creator with certain unalienable Rights, that among these are Life, Liberty and the pursuit of Happiness."

So, if the Declaration of Independence does not guarantee happiness, why does this procedure always end on "Happiness" regardless of the initial word you pick?

A column that might strike a few people, perhaps some QAnon members, as amazing.

WEIRD SCIENCE FROM THE HUME AND BACON INSTITUTE

While surfing the internet, I chanced on the Hume and Bacon Institute for Alternative Science just outside Edinburgh, Scotland.

This low-profile research facility has just published some of its recent findings in the British science journal *Languet* (vol. 233, 1999). As the details of the article become known, they will prove quite startling to many.

In a news conference, Paul A. Yiannis, the institute's shy yet pugnacious director, stated that his staff had convincingly verified a number of controversial claims. UFOs and ESP.

The first involves a strange piece of metal found in Roswell, New Mexico, where many believe an alien spacecraft crashed in 1947.

The fragment has quite an amazing property: It exerts a faint physical attraction on every information-processing instrument so far tested. Moreover, this attraction is nine times as strong one foot away from the metal as it is one yard away.

What to make of the fragment's effect is open to differing interpretations, but it can no longer be denied. Yiannis maintains that the relation of this phenomenon to the new physics of string theory will be an active area of research in the 21st century.

The institute's article also deals with psychic readings. It concludes that labs all over the globe have demonstrated that psychics are indeed frequently correct in their assessments of others.

Moreover, there are documented cases where they have correctly described the characteristics and life experiences of dead relatives of subjects. Yiannis also pointed to the literally thousands of cases of dreams prefiguring events that occurred the next day and mentioned the counterintuitive laws of quantum mechanics. Trying to determine the consequences of these laws, he referred to puckishly as "walking the Planck," after the German physicist Max Planck.

Toward the end of the news conference, Yiannis showily popped a pill and enthused, "And while I'm at it, I can state with certainty that the taking of homeopathic remedies and countless other alternative remedies is followed many times by complete remission. This is especially so after the patient has had a serious medical crisis. The dismissal of these claims of remission would be the act of a closed mind."

Again, the paper contains copious documentation and personal testimonials. The institute's 83-page *Languet* survey of alternative science also refers to a report on the trigonometry and calculus that practitioners need for astrological and biorhythmic predictions. Sometimes, it's considerable, and the claim is made that many equations and theorems used in astrology and biorhythmic analysis are universal truths.

Related to these phenomena, there's a brief discussion of the strange numerological properties of the numbers 23 and 28. The numbers, some believe, are the periods for a metaphysical male and a female principle, respectively. They have the special property that by adding and subtracting appropriate multiples of them, one can express any whole number whatsoever. That is, any number at all can be written as $23X + 28Y$ for suitable choices of X and Y. For example, $6 = (23 \times 10) + (28 \times -8)$.

The institute's paper is evasive on some issues as, for example, when it underscores that UFOs are indeed unidentified flying objects but doesn't address the issue of alien abductions. And some areas of alternative science are deemed invalid by the Institute's staff.

One of these is phrenology, the belief that head shape helps determine personality. Yiannis, however, seemed a bit self-conscious when answering a reporter's question on phrenology. Onlookers couldn't help but notice his small, lumpy, asymmetric head.

The reaction of the scientific community to Yiannis's claims in *Languet* has been muted so far. I, for one, endorse his assertions—with a few qualifications listed below.

No doubt, many of you have already realized that Paul A. Yiannis's pronouncements are all true—and all misleading. ("Yiannis" is Greek for "John," making the director's name the rough reversal of mine.)

Let's look at the assertions made above:

Any two objects attract each other with a force inversely proportional to the square of the distance between them, which is what the foot–yard attraction differential amounts to. This is simply Newton's law of gravitation, and its relevance is on a par with that of other laughable "evidence" from Roswell.

Of course, psychics are frequently correct. It would be truly mind boggling if they were always wrong. What is not true, however, is that they are correct at a rate higher than that of chance.

Yes, quantum mechanical results are weird. Sprinkling a little quantum hocus-pocus about clarifies nothing, however, although it does tend to impress those who like their science incomprehensible.

Remission does often follow the taking of homeopathic and other remedies, especially after a crisis when getting better is not at all unlikely. Improvement may even follow putting mustard in one's left ear. Again, each proposed remedy must be shown to bring about a higher rate of remission and improvement than the norm; it's not sufficient to show that it sometimes "works."

Trigonometry and calculus are flawless and indubitable; it's just that the assumptions that underlie their use in astrology and biorhythms are totally bogus.

Any whole number can be written as $23X + 28Y$, but that is because 23 and 28 are relatively prime—they have no factors in common. Any other relatively prime numbers, say, 15 and 22, have the same property, which is devoid of mystical significance.

UFOs are just that, unidentified flying objects, and the world is full of objects that have not been identified—on the ground, in the air, under the sea, even in the mote-filled air under one's bed. This doesn't mean there are aliens intent on abduction or nano-robots planning to colonize our bloodstreams.

Chapter Four

Calculations and Miscalculations

Not everything that can be counted counts, and not everything that counts can be counted.

—Albert Einstein

Too many numbers and statistics, even when well supported by evidence, are presented baldly and without context and are thus liable to misinterpretation. A favorite example of mine, among countless others, is the five-year survival rate as a measure of the effectiveness of a treatment for a disease. This is, of course, an important measure, but as I've noted elsewhere, it has its flaws, especially when there is no really good treatment for the disease in question. Consider, for an extreme case, a particularly virulent cancer for which there is a screening test. There are, let's also assume, two states, one of which mandates early screening for a disease and has a 100% five-year survival rate, the other of which does not believe in screening for this disease and has a 0% five-year survival rate. Which state will have fewer deaths from the disease in question is pretty clear-cut, right? Possibly not.

To illustrate, suppose that the disease is such that it arises only when people are in their early 50s and always proves fatal, with or without treatment, at about the age of 65. (Clearly, the numbers in this example are made up, but they're sufficiently similar to the real numbers for some diseases to be instructive.) Further, let's suppose that clinical symptoms for this disease don't manifest themselves until people are in their early 60s. The state that mandates early screening when people turn 50 will, given the above assumptions, have a 100% five-year survival rate, while the state that discourages such screening will have a 0% five-year survival rate. Yet all people who develop the disease will die at about 65.

Assuming more realistic numbers and a treatment that is not particularly protective will lead to a similar failure of five-year survival rates as a measure of the effectiveness of a treatment. This is not intended to minimize the importance of screening but simply to point out that it, like most metrics, doesn't always measure what it purports to measure.

The normal bell-shaped (and other) statistical distributions are also inappropriate metrics in many situations. One reason is that their very shape can give rise to flawed conclusions and cocksure policy analysts. This is because even a small divergence between the averages of different population groups is accentuated at the extreme ends of these curves, and these extremes, as a result, often receive a lot of attention, some of it wrongheaded. (To see this, imagine two bell-shaped graphs on the same axis. When one is pulled slightly to the right, the extreme ends or tails of the two graphs are exaggerated.)

More specifically, consider two population groups, say, of men in their 20s, perhaps from different countries or different ethnicities, and that each of the two groups vary along some dimension—systolic blood pressure, for example. Let's also assume that each of the two groups' systolic blood pressures vary in a normal or bell-shaped manner with similar variability within each group. Then, even if the average blood pressure of one group is only slightly greater, say, 122, than the average blood pressure of the other, 118, people from the first group will make up a significant majority of those with very high blood pressure.

Likewise, people from the second group will make up a significant majority of those with very low blood pressure. This is true even though the bulk of the men from both groups have approximately normal systolic blood pressure. Of course, blood pressure isn't at all an emotionally evocative issue, yet the same dynamic operates if the issue does arouse passions. Yet another factor is that groups in a population not only differ along some dimension but also occupy differing percentages of the whole population. They differ as well in the extent to which they overlap with each other.

The bottom line is that we should refrain from automatically attributing to racism what might simply be a deviation from statistical uniformity for a variety of other reasons.

Flaws of this and other sorts hold with a variety of measures. The numbers given and conclusions drawn are sometimes tenuously related to what it is they're supposed to demonstrate. Crime statistics, COVID-related figures, economic forecasts, SAT scores, relationship questions, and sex trafficking figures are some examples considered below.

Even accurate numbers can be misleading. For example, the recent forest fires in the West as of this writing have burned about a million acres. Since there are 640 acres per square mile, this figure is approximately equivalent to a 40-mile × 40-mile square, a formulation that seems considerably less alarming although no less terrible. Expressing the area in terms of acres rather than square miles brings to mind the taunt, "Liar, liar, pants on fire." I certainly don't mean to minimize the existential threat to the climate that is being greatly exacerbated by global warming but merely to point out the need for more clarity whenever possible.

One last example of naked data without the crucial background. The TV sportscaster announces, "Folks, we're running out of time, but I want to squeeze in the baseball scores: 5 to 1, 7 to 4, 9 to 6, and in extra innings, 3 to 2."

A tale of two numbers that differ radically in political importance. Which one received more attention? And which one is still a source of contention?

From Dates and Y2K to People and PY2K: A Tale of Two Numbers

The year 2000 and the U.S. population as determined by the next U.S. Census: One number is captivating, written about everywhere, and basically unimportant, while the other is boring, written about only sporadically, and quite important.

First, the millennium. As has been often observed but more often ignored, the third millennium begins on January 1, 2001, not January 1,

2000. President Clinton's grand bridge to the new millennium is technically just a small overpass to the last year of the 20th century. The explanation is that in contrast to the Y2K problem, the solution to the earlier Y0K problem (i.e., the transition from BC to AD) was constructed long after the fact and stipulated that the dating system would begin with a year 1 and not a year 0. Such purity aside, the numerological appeal of 2000 is clearly much greater than that of 2001 (except to Arthur C. Clarke). Its only social importance, however, is that it gives us all a sense of commonality derived ultimately from the fact that we have 10 fingers, the origin of our base-10 number system. If we had eight fingers and consequently had developed a base eight numbering system, the year **2000** (which is short for $2 \times 10^3 + 0 \times 10^2 + 0 \times 10^1 + 0 \times 1$) would be written **3720** (short for $3 \times 8^3 + 7 \times 8^2 + 2 \times 8^1 + 0 \times 1$), a much less impressive sequence of digits.

Talk of the new millennium also brings to mind apocalypse and the nonsense surrounding the number 666, which would not have come about in, say, a base-5 world. Had we counted on the fingers of only one hand, the odious **666** would have been innocuously expressed as **10131** ($1 \times 5^4 + 0 \times 5^3 + 1 \times 5^2 + 3 \times 5^1 + 1 \times 5^0$). By an application of the principle of conservation of superstition, however, in such a world, 10131 might have been referred to as the Sign of the Beast, equipped with a silly backstory about the three 1s and the one 3 in it.

But, alas, we live in a base-10 world, and thus we are all committed decacrats, and this brings me to a less universal party, the Democrats, and the other big number that looms before us: the U.S. population in the year 2000. We may perhaps dub this the PY2K problem. In contrast to the arbitrary precision and global scope of the millennial year, U.S. Census figures are necessarily imprecise and of more parochial interest. Nevertheless, they are politically and economically significant; legislative apportionment, economic assistance, business decisions, and much else depend on them. Moreover, and this is the connection with the Democrats, when sampling techniques are used to estimate the uncounted, the folks found tend to be members of minorities and to vote Democratic.

Unfortunately, the U.S. Supreme Court has just ruled that statistical sampling techniques for purposes of federal legislative apportionment are forbidden by the U.S. Constitution, and so we're limited to a simple (and incomplete) head count and a census not nearly as accurate or fair as it could be. It's easy to parody the hidebound nature of the decision. If census workers know that each of 50 tracts contain 200 citizens, are they permitted to use high-tech multiplication to conclude that 10,000 people live in the region, or would they have to limit themselves to an "actual enumeration"? Would the Bureau of Standards content itself with using the length of Benjamin Franklin's forearm as its unit of measurement? Will there eventually be two sets of figures: the official count and the real although unofficial one, supplemented, where needed, by statistical estimates of the uncounted?

Whether they're of the ethereal, mystical sort like 2000 and 666 or of a grubbier but more important variety like the census figures, numbers have long helped define (and sometimes misdefine) us. Pythagoras was certainly ahead of his time when he proclaimed, "All is number" around 500 BC (or was it –500 AD?).

Exercise 1: Show that in base 7, the year 2000 would be written as 5555.

Exercise 2: Harder problem. Estimate how many congressional office seekers opposed to the use of statistical techniques for the U.S. Census will refrain from hiring pollsters to sample their own constituencies.

Solutions:

Exercise 1: Show that in base 7, the year 2000 would be written as 5555. In base 7, **5555** indicates the number $5 \times 7^3 + 5 \times 7^2 + 5 \times 7^1 + 5 \times 1$, which sums to 2000.

Exercise 2: Harder problem. Estimate how many congressional office seekers opposed to the use of statistical techniques for the U.S.

Census will refrain from hiring pollsters to sample their own constituencies. Excepting candidates from safe districts, I would guess that
the number forgoing the services of pollsters is quite small.

We can use spherical cows or rough estimation techniques to illustrate
an instance of apophenia, the seeing of patterns that aren't there but that
are often prompted by fear. COVID vaccines and blood clots arouse basic
emotions, so many of the common and baseless reactions to the vaccine
and its risks unfortunately are to be expected.

Apophenia and Spherical Cows: Clots and Coincidences
First off, let me say that a very small handful of the blood clots that
occurred after the COVID vaccine was administered were found to be
very rare, very deadly, and very likely precipitated by the vaccine.

I know nothing about these few cases, but a back-of-the-envelope
calculation suggests that many people will suffer garden-variety blood
clots within a week of being vaccinated, and unfortunately a good number of them will attribute them to the vaccine and not to coincidence.
Why is this very likely?

In the spirit of physicists' common use of the phrase "spherical cow"
to refer to simplified models that are stripped of realistic details for ease
of computation and proof of concept, let me explain by first making some
simplifying assumptions about the relevant numbers and time intervals.

Let's assume unrealistically (since this is just an exercise in rough
estimation) that 250,000,000 people will be vaccinated within a year at
an assumed average uniform rate of 5,000,000 per week. (250,000,000/52
is about 5,000,000.) The inferences made below hold regardless of the
real numbers. We do know there are approximately 800,000 blood clots
leading to 80,000 deaths, most of them women, each year in the United
States. Assuming an average rate, this reduces to a bit more than 16,000
blood clots per week. (800,000/52 is about 16,000.) (Let me focus on
blood clots, although a similar analysis applies to heart attacks and many
other conditions.)

Given these spherical cow–like assumptions, let's examine a particular week in which a group of 5,000,000 people are vaccinated. How many of the 16,000 cases of blood clots will happen to these 5,000,000 people during this particular week? About the fraction 5,000,000/330,000,000 (the denominator is the population of the United States) times 16,000 people, which equals about 250 people who will suffer serious clots sometime during the week after they've been vaccinated. Thus, there will be 250 (again just a rough ballpark estimate) of the 5,000,000 vaccinated that particular week who will have serious clots within the week after their vaccination, some of them fatal.

Arguably, one could also say that there'll be about 250 such clots within a week after vaccination during every week of the next year, approximately 12,500 for the year. These would be coincidental, but given the perennial problem of American innumeracy and the present political environment in which random anecdotes and baseless tweets often trump reams of research, it seems certain that many people will ascribe significance to these coincidences. No doubt, outlandish explanations for them will be constructed to explain what needs no explanation. Making matters worse is that similar remarks would apply to other diseases and conditions as well.

As with bovine sphericity ("vaccine" incidentally derives from the Latin word for cow), all of the numbers and assumptions above could and should be challenged. I have refinements for and reservations about them myself, but even if the above estimate is off by a significant margin, it would still be quite enough to induce an unwarranted high degree of anxiety.

The calculation above serves to show that we should anticipate many cases of a seeming causality between vaccines and blood clots (or many other naturally occurring conditions). Apophenia, seeing patterns where none exist, will be ubiquitous especially given the life-or-death stakes involved. Moreover, it's very likely that if this perception is unchallenged, it will continue to lead many to refuse vaccination. That would be a shame for them and for the public at large.

Except to the occupationally myopic such as myself, calculus and COVID at first glance seem a strange pairing. Nevertheless, linking incongruous subjects is often worthwhile. Here I simply point out to those who know a bit of calculus that looking for the point at which the COVID curve flattens is quite distinct from searching for its inflection point. I also point out what the area under the curve represents.

COVID, Calculus, and the Curve

The novel Coronavirus has generated an overwhelming and confusing torrent of science, pseudoscience, and pseudomath on the internet, much of it relevant to other diseases. One statistic, however, is compelling, and that is that with about 4% of the world's population, the United States has been home to about 25% of the world's COVID-19 deaths, a consequence of the Trump administration's catastrophic incompetence. (As of this later writing, the percentage is somewhat lower.)

There has been a lot of real mathematics implicit in the reporting on the disease. The ubiquitous COVID-19 curve has been a particular focus of interest, so let me make a few observations about it using quite intuitive notions from calculus. Commentators online, in print, and on television incessantly discussed flattening it, but no pundit to my knowledge invoked ideas from the subject to describe it. Don't worry. No formal study of calculus is needed to understand the gist of the following, but those who have studied the subject may especially appreciate its use here.

The x-axis of the curve is the increasing passage of time in days, and the y-axis is the daily death toll from the disease. Generally, the curve rises slowly but grows exponentially. After a while, the growth continues but at a slower rate and eventually levels off and starts to decline.

Ideally, of course, we want this leveling off or flattening of the curve to occur sooner rather than later.

To use notions from calculus, let's abuse the spiky curve for a bit and deal with a smoothed-out version of it whose graph can be described by a formula or function, $f(t)$, that depends on the time. At the curve's peak, it is of course flat. In mathematical terms, we say that the first derivative of the function, indicated by $f'(t)$, is zero. This means that the rate of change of the daily death toll is zero there. It is neither increasing nor decreasing there.

Another special point on the graph of f(t) is where the daily death toll is still growing but beginning to grow more slowly. This is termed an inflection point of the graph, and mathematically it is a point where the second derivative of the function, indicated by f " (t), is zero. This means that the rate of increase of the daily death toll, though positive, is beginning to slow.

Most growth reaches an inflection point sooner or later. Consider the height of a child. A very young child grows very rapidly until a certain age, and then as the child grows older, its growth rate, while still positive, slows down. The point at which this happens is an inflection point and is to be distinguished from the maximum or peak adult height where the growth levels out and stops. Although sometimes confused, the peak and the inflection point are generally two distinct points.

Another observation about the curve is that the area underneath it gives the total number of deaths up to a point in time. To see this, imagine that we replace the changing number of deaths per day with the changing velocity in miles per hour of a moving car throughout time. In general, the area under such a curve would give the total number of miles traveled up to a point in time. (Say, for example, a graph of the car's velocity showed that it traveled 30 miles per hour for two hours and then 50 miles per hour for three hours. The total distance traveled under this graph would then be 210 miles.) We can now bring out a big gun from calculus and invoke the fundamental theorem to conclude that the area under the COVID-19 curve is in fact the number of deaths up to a point in time just as the area under a velocity curve is in fact the number of miles traveled up to a point in time.

A last rather depressing note: There is no good reason to suspect that the curve describing the daily death toll has just one peak. The toll of the virus will likely rise and fall several times before the virus is vanquished, or it might just stay at a horrific plateau.

As was the case with Cipro and the anthrax scare, an issue involving the extended prisoner's dilemma arose with masks and COVID. To cooperate and do what's best for the community (wear masks and not hoard Cipro, a minimal social obligation) or to defect (not wear a mask and hoard Cipro, arguably a personal benefit), that is the choice.

Innumeracy also was an essential ingredient in the irrational response to both issues. Big numbers and small numbers didn't mean anything to many people. They merely serve as a kind of decoration. Thus, the impact of extremely rare events in both the Cipro and the COVID situations—a handful of cases of a most atypical sort of blood clot and a comparably small collection of anthrax-laden letters—was unfortunate but not surprising.

The response to global warming and fossil fuels is also an example of defecting and going all in on fossil fuels or cooperating and quickly weaning ourselves from them. Here too, the numbers and the science don't seem to elicit an adequate recognition of this existential threat. The relevance of this shortsightedness to greenhouse gases, extinct species, and global warming should be obvious. Environmental despoliation may even be conceived of as a kind of global Ponzi scheme, the early "investors" (i.e., us) doing well, the later ones losing everything. Hard not to Ponzificate on this.

HOARDING MEDICINES TO WEARING MASKS: SOCIOLOGY, PUBLIC HEALTH, AND THE PRISONER'S DILEMMA

One of the fundamental insights of the eminent sociologist Emile Durkheim, arguably the founder of sociology, was that social "laws" are not necessarily derivable from psychological ones. Even the seemingly most personal decisions Durkheim writes, such as whether to commit suicide, can be explained in part by sociological factors that are external to the individual and part of the ambient culture that conditions our actions and is necessary for a society to cohere.

There is often a disconnect, however, between society and the individual, and these disconnects can lead to dilemmas. Often, these arise when social constraints make people feel powerless and vulnerable. A case in point is the situation regarding the drug Ciprofloxacin, which some are stockpiling more to combat anxiety than to ward off anthrax. The benefit of these purchases was a feeling of greater personal security, but one social cost is that Cipro may be in short supply if and when it's needed in large quantities. Another social cost is the increased bacterial resistance to this antibiotic and others that is likely to result from our widespread use.

The verbs in the previous paragraph are past tense, but the phenomenon is present tense as the various COVID-19 restrictions illustrate.

The Prisoner's Dilemma

The so-called prisoner's dilemma is often used to model such conflicts. The dilemma owes its name to the scenario wherein two men suspected of a burglary are arrested in the course of committing some minor offense. They're separated and interrogated, and each is given the choice of confessing to the burglary and implicating his partner or remaining silent.

If both remain silent, they'll each receive only one year in prison. If one confesses and the other doesn't, the one who confesses will be rewarded by being let go, while the other one will receive a five-year term. If both confess, they can both expect to spend three years in prison. The cooperative option is thus to remain silent, while the individualist option is to confess.

The dilemma is that what's best for them as a pair, to remain silent and spend a year in prison, leaves each of them open to the worst possibility: being a patsy and spending five years in prison. As a result, they'll probably both confess and both spend three years in prison.

Most of us aren't crime suspects, but the dilemma provides the logical skeleton for many situations we do face every day in real life. Whether we're competitors conducting business, spouses negotiating understandings, or anxious citizens vying for antibiotics, our choices can often be phrased in terms of the prisoner's dilemma. The two parties involved will often be better off as a pair if each resists the temptation to go it alone and instead cooperates with and remains loyal to the other person. Both parties' pursuing their own interests exclusively leads to a worse outcome than does cooperation.

This two-person prisoner's dilemma situation can be extended to many people, each having the choice whether to make a very small contribution to the public good or a much larger one to his or her own private gain.

These small contributions add up, however, and society as a whole is better if more people take the cooperative option. This many-party prisoner's dilemma, useful in dealing with environmental goods such as clean air and water, is most relevant to the situation regarding Cipro.

If we refrain from buying our own supplies of Cipro, there will be more available in any emergency, and the bacteria that constitute our common environment will not have as many tutorials to help them learn how to outwit the antibiotics.

Minuscule Risk

Alas, this is not to say that buying Cipro in anticipation of a possible emergency never makes sense, especially if one believes that the conditions of the prisoner's dilemma simply do not apply or that the public health system will not be up to the job in an emergency.

The best way public health people can minimize hoarding is to repeatedly stress that there is not yet such an emergency. And short of an unpredictable and improbable scientific breakthrough by a scientist in the employ of terrorists, the risk from anthrax is tiny.

Nearly 700 million pieces of mail are delivered daily, and there have been only a handful of cases, almost all of them treated successfully. By contrast, approximately three-quarters of a million Americans die annually from heart and circulatory diseases, around half a million from various forms of cancer, and more than 35,000 in car accidents. Even the much-derided and now almost idyllic-seeming shark menace has resulted in more deaths.

(Incidentally, a positive spin on the anthrax-laden letters is that they may indicate that the perpetrators don't have anything more virulent. Why would they warn us with isolated missives if they were capable of something much more horrific? The terrorists who attacked the World Trade Center did not first crash small Cessna planes into the towers as a prelude to doing so with jetliners.)

Another way to limit private stockpiling is for authorities, preferably scientists rather than politicians, to clearly proclaim that penicillin and doxycycline are also effective in combating anthrax and that there is no shortage of these drugs. Finally, if and when much more Cipro is deemed necessary, government officials can always break the drug's patent, as Canada has done prematurely, and go to generic versions of the drug.

The bottom line is that private stockpiling of antibiotics makes no sense for most people. Nevertheless, for the relatively few who feel especially vulnerable because of their psychology, physical location, or occupation, buying the drugs is not an irrational way to increase their feeling of security (as long as they refrain from taking them without a very good reason to suspect exposure).

The hysteria generated by the few anthrax-laden letters is dangerous and counterproductive. Resisting it is almost a patriotic duty, and anything that helps to do so is probably a good thing.

Once again, the comparisons with COVID are most appropriate.

The "butterfly effect" is the idea that the flapping wings of a butterfly could months later influence the weather thousands of miles away. Economics is likely to be influenced by such tiny effects. So, as this piece showed, might ants.

OF ANTS, BUTTERFLIES, AND ECONOMIC WHIMSY
The deadline is tomorrow for a proposal I promised to deliver but now don't want to write. I dutifully, if reluctantly, begin to work on the project when a niggling detail about some utterly irrelevant matter comes to mind. It may concern the etymology of a word, the colleague whose paper bag ripped open at a departmental meeting revealing an embarrassing magazine inside, or why caller ID misidentified a friend's telephone number. These in turn bring to mind the next in a train of associations and musings.

Such prosaic episodes strongly suggest to me that there will never be a precise science of economics. Shopping and buying, I suspect, sometimes partake of a similar whimsicality.

I was thus fascinated by a new book by Paul Ormerod, a British economic theorist and professor who has worked for *Economist* magazine. Titled *Butterfly Economics*, the book faults the discipline for not sufficiently taking into account the commonsense fact that people influence each other in myriad subtle ways.

People do not, as orthodox economics maintains, have a set of fixed preferences that they coolly and rationally base their economic decisions on. The assumption that people are sensitive only to price simplifies the mathematical models, but it is not true to our experience of fads, fashions, and everyday "monkey-see, monkey-doism."

An experiment involving not monkeys but ants provides a guiding metaphor for the book. Two identical piles of food are set up at equal distances from a large nest of ants. Each pile is automatically replenished, and the ants have no reason to prefer one pile to another.

Entomologists tell us that once an ant has found food, it usually returns to the same pile and that on coming back to the nest, it physically stimulates other ants to follow it to the same pile.

So where do the ants go? It might be speculated that either they would split into two roughly even groups or perhaps a large majority would arbitrarily settle on one or the other pile. The actual ant behavior is counterintuitive. The number of ants going to each pile fluctuates wildly and doesn't ever settle down. Graphing these fluctuations results in something that looks suspiciously like the variations in the stock market.

And in a way, the ants are like stock traders. On leaving the nest, each ant must make a decision: go to the pile visited last time, be influenced by a returning ant to switch piles, or switch piles of its own volition. This slight openness to the influence of other ants is enough to ensure the complicated and volatile fluctuations in the number of ants visiting the two sites.

Ormerod takes off from here to discuss Christmas toy fads, surprise movie hits and misses, the triumph of inferior VHS machines over Betamaxes in the VCR market, crime and divorce rates, and, most extensively, economics—all phenomena that derive, in part, from each of us influencing and being influenced by others.

The book's title refers to the notion that a butterfly flapping its wings in, say, Venezuela, might spell the difference several months later between a hurricane and a balmy day along the eastern U.S. seaboard. The book's thesis is that interactions among economic agents sometimes introduce nonlinear effects (small causes, disproportionately large consequences) that make long-range precise predictions all but impossible. This is most certainly not to say that rough, qualitative predictions or even more precise, quantitative ones are impossible.

Governments, says Ormerod, "should do very much less in terms of detailed, short-term intervention" but rather should search for the broader patterns and not respond to ephemeral and short-lived shifts in supply and demand or human tastes and preferences.

Not only do most orthodox economists take too little notice of the influence we exert on each other, but many don't sufficiently appreciate the unpredictable consequences of the interconnectedness of economic variables. Interest rates have an impact on unemployment rates, which in turn influence revenues; budget deficits affect trade deficits, which sway interest rates and exchange rates; an increase in some quantity or index positively (or negatively) feeds back on another, reinforcing (weakening) it and being in turn reinforced (weakened) by it.

I wrote about using nonlinear dynamical systems to model such interconnections in *A Mathematician Reads the Newspaper*. For illustration, I described a simple instance of one: Imagine that approximately 30 round obstacles are securely fastened to the surface of a pool table in haphazard placement. Hire the best pool player you can find and ask him or her to place the ball on the table and take a shot toward one of the round obstacles. Then ask for the exact same shot from the exact same spot with another ball. Even if the angle on the second shot is off by the merest fraction of a degree, the trajectories of these two balls will very soon diverge considerably, magnified by each succeeding bounce. Soon, one of the trajectories will hit an obstacle that the other misses entirely, at which point all similarity between the two trajectories ends.

The butterfly effect, the sensitive dependence on initial conditions, here takes the form of the sensitivity of the billiard balls' paths to minuscule variations in their initial angles. It's not totally unlike, say, the dependence of one's genetics on which zigzagging sperm cell reaches the egg first.

Consider also the disproportionate effect of seemingly inconsequential events: the missed planes, serendipitous meetings, and odd mistakes that shape and reshape our lives.

Human interactions and interconnected economic variables strongly suggest that the economy is less susceptible to precise, long-range forecasts than many believe.

Staying with the butterfly effect, I believe that fuzzy imprecise numbers are usually the only kind of numbers that are appropriate when it comes to describing or explaining the economy. Such descriptions would benefit from the citing of confidence intervals. Precision is regularly claimed by politicians and others, but we often should react to such claims as we would to someone who announces that he weighed 168.216923 pounds at 8:05 in the morning. My apologies for my perhaps excessive focus on this bugbear of mine.

THERE'S NOTHING WRONG WITH FUZZY MATH

It's a safe bet that you have never heard a presidential candidate explain how adjusted multiple correlation coefficients work. (And that's a very good thing too.)

Instead, they talk a lot about individuals with specific problems (89-year-old Mrs. Kadoskins from rural Nebraska, whose prescription bill is exorbitant) that, it is hoped, will resonate with millions of other people.

Politicians tell stories and anecdotes because they're much easier to understand than numbers. Statistics, if they're more complicated than batting averages, usually go relatively unnoticed. And even when politicians do insist on citing the relevant figures, they're liable to be mocked. In the first debate, George W. Bush spoke derisively of Al Gore's "fuzzy numbers."

As a fuzzy mathematician, I took umbrage. What's wrong with fuzzy numbers? It is my opinion that, in politics or economics, there are no other kinds of numbers. Polls explicitly acknowledge some of their fuzziness by including margins of error, but fuzziness is implicit in many other contexts. Repeating the same point about precision, I note that if a politician were to say that the size of his tax cut was exactly $1,265,155,844,138.36, he would surely be delusional.

Economic numbers with a couple of significant digits (say, $78.3 billion) are usually the most we can hope for, but even this is often impossible due to all sorts of wild cards. A big unknown today is whether the current budget surpluses will last. Both candidates are banking on them, but that's certainly no guarantee.

Consider the complex interactions among interest rates, unemployment, revenues, budget and trade deficits (or surpluses), stock prices, and so on and dizzyingly on.

Another source of fuzziness is that one can define the same quantity in many different ways. Is Bush giving away more in tax cuts to the richest 1% of Americans than he will spend on a variety of social programs? If the estate-tax repeal is included, the answer is yes. If the fact that the very rich bear roughly 30% of the tax burden is given too much weight, the cuts may be considered less outrageous and the phrase "giving away" less apt.

Yet another impediment to precision derives from the mathematical discipline usually referred to as chaos theory. The associated notion of a nonlinear dynamical system is quite relevant to these economic complexities. There are a number of simple illustrations of such systems involving the aforementioned balls on a pool table or water on a spinning wheel, but most such real-world systems are much more complicated. Nevertheless, the simple illustrations provide at least a vague, intuitive understanding of the effects of many nonlinearly interacting variables. These effects should be sufficient to arouse a certain wariness of simplistic pronouncements delivered with overweening confidence.

Nonpartisan experts regularly disagree on the numbers. A recent issue of the *Economist* contained a very revealing article in which the prestigious magazine asked 54 eminent economists to grade the Bush and Gore economic plans. Their varied responses reminded me of President Truman's quip about the desirability of a one-armed economist who would be unable to waffle and say, "But on the other hand. . . ."

The most interesting result of the *Economist* survey, however, was the uncertainty and wide range of expert opinion on most aspects of the candidates' economic plans.

Given these well-reasoned but different opinions, I wonder why the candidates so seldom say anything remotely like "I don't really know," "I'm not at all certain," or, simply, "Beats me." Surely, this sort of admission is the correct response to so many of the economic questions put to Gore and Bush by Jim Lehrer that its nonutterance should be something of an intellectual scandal. It isn't.

Alas, one reason for politicians' pose of confidence and surety is not hard to fathom. They fear that voters will confuse a slow, qualified response with ignorance or evasiveness, while they hope voters will confuse a quick, glib response with knowledge or resolve. As a result, candidates feel they must sometimes square their jaws and forthrightly make pronouncements that are the political equivalent of predictions from the Psychic Network.

But might the electorate not be more impressed by an occasional and courageous confession of ignorance—not ignorance of basic facts but of the economic consequences of adopting a particular policy? Certainly, economic history, the mathematical discipline of chaos theory, and common sense strongly suggest that uncertainty and tentativeness frequently are more than justified.

Of course, I could be wrong. My electoral accomplishments have been limited to a losing campaign for high school senior class treasurer.

Studies have shown that the correlation between SAT scores and first-year college grades is not overwhelming. Relativizing the scores to the colleges to which students apply, however, results in a somewhat stronger correlation. Still, there's much that the scores don't come close to measuring as the other lesser-known SAT, the Soccer Assessment Test, suggests.

PREDICTING SUCCESS? SAT SCORES AND COLLEGE GRADES AND THE OTHER SAT, THE SOCCER ASSESSMENT TEST

School's getting out, but the dreaded Scholastic Assessment Test, better known as the SAT, looms just a summer away for next year's high school seniors.

Given this, many might be inclined to agree with the president of the University of California, who announced several months ago that he would like to abolish the test as a requirement for admission to the school. (He would retain the SAT II, which measures achievement within particular disciplines.)

The announcement of his intention sparked an ongoing controversy. The issue is complicated, and any full discussion should address the issue of the change in scores throughout the years. (I wish I had a dollar for every baby boomer I've heard say that the scores started to decline just after they graduated from high school.)

Other important issues are the renorming of the test, culturally biased questions, ethnic and gender differences in the scores, self-selection of test takers, differential participation of various subgroups, the importance of calculators, test preparation, certainly the economic status of the family, and a variety of other personal and cultural factors.

The big question, however, is, How predictive of success in college are SAT scores? More precisely, what is the correlation between high school SAT scores and first-year college grade-point average (GPA)? (The appropriateness of GPA as a measure of success or innate talent is open to question. Grades, for example, often depend critically on the courses taken.)

Most studies find that the correlation between SAT scores and first-year college grades is not overwhelming and that only 10% to 20% of the variation in first-year GPA is explained by SAT scores. This association appears weaker than it is, however, for an interesting but seldom noted statistical reason: Colleges usually accept students from a fairly narrow swath of the SAT spectrum.

The SAT scores of students at elite schools, say, are considerably higher, on average, than those of students at community colleges, yet both sets of students probably have roughly similar college grade distributions at their respective institutions.

If both sets of students were admitted to elite schools or both sets attended community colleges, there might well be a considerably stronger correlation between SATs and college grades at these schools. Those schools that attract students with a wide range of SAT scores generally have higher correlations between the scores and first-year grades.

This is a general phenomenon; the degree of correlation between two variables depends on the range of the variables considered.

Soccer Assessment Test (SAT)

The SAT traditionally deals with analogies, so let's consider soccer leagues. Assume there were an SAT (Soccer Assessment Test) that measured the speed, coordination, strength, and soccer experience of students in a certain city. Assume further that the students roughly divided themselves into five leagues depending on their scores on this SAT, players in the top leagues having higher SAT scores on average than those in the lower leagues.

One wouldn't expect that a measure of success in the sport, say, number of goals scored, to vary much among the leagues. There would be good scorers and bad scorers in every league, and, just as grade-point distributions are similar in most colleges, the distribution of goals scored would probably be similar in the five leagues.

In each league, the better scorers would probably have only slightly higher SATs on average. In other words, there wouldn't be a high correlation between SAT scores and success in soccer within any league. There would, however, likely be a much higher correlation between SAT scores and soccer success were the students randomly assigned to the teams in the five leagues. (Similar remarks could be made about boxing, the different weight divisions washing out much of the correlation between greater weight and success at boxing.)

Of course, there are many dimensions of soccer ability that aren't measured by this imaginary SAT just as there are many, many dimensions of scholastic ability that aren't measured by the SAT. Concentrated work for an extended period is certainly one of the latter, the premium the SAT places on one morning's speedy work being especially difficult to defend. There is much else the SAT fails to measure.

The analogy between soccer and scholastics is not perfect, of course, but the point remains. Like the soccer SAT, the scholastic SAT provides incomplete but useful information to students and colleges. A rough measure of intellectual preparedness, the SAT shouldn't be made into a fetish, but it is not clear that it should be ignored.

Without it, colleges would undoubtedly place more emphasis on high school grades and extracurricular activities, measures that also have serious shortcomings—grade inflation and meaningless résumé puffing being the main ones. The SAT is a flawed predictor, but it is also relatively objective and, among other virtues, sometimes provides a way for the bright yet socially inept student to be recognized.

I might add that it is unfortunate that family income is a very good predictor of success along a variety of dimensions.

A Tricky Problem

A mathematics question is not inappropriate in a discussion of the SAT, so let me include here a tricky one. In the correctly solved additions below, each of the five letters represents a different digit, EA being a two-digit number. What is the value of B + D if

A C

B␣D

C EA?

Solution: Combining the two additions yields A + B + C + D = C + EA. If we cancel the Cs from both sides of this equation, we obtain A + B + D = EA, and thus B + D = EA − A. The two-digit number EA of course equals $10 \times E + A$, and so EA − A equals $(10 \times E + A) - A$, or simply $10 \times E$. Since the digit E must be 1, (B + D) = 10×1, or just plain 10. There are other approaches as well.

Person 1: Artificial intelligence and machine learning algorithms eliminate bias and human fallibility.

Person 2: Really, I'm a bit dubious. Where do these algorithms come from?

Person 1: From reams of data gleaned from newspapers, magazines, books, movies, and so on.

Person 2: I rest my case.

BE CAREFUL WHAT YOU MEASURE AND TRY TO ACHIEVE—YOU MIGHT SUCCEED

Let me mention a most apt illustration of the extra-mathematical assumptions necessary in many applications of mathematics. Two men often credited with the beginning of probability theory, Pierre Fermat and Blaise Pascal, are said to have bet on a series of coin flips. They agreed that the first to win 6 such flips would be awarded $1,000. The game, however, was interrupted after only 8 flips with the first man leading 5 heads to 3 tails. The question is, How should the pot be divided? It might reasonably be argued that the first man should be awarded the full $1,000 since it was implicit that the bet was all or nothing and he was leading. An alternative view might well be that the first man should receive 5/8 of the pot and the second man the remaining 3/8 since the score was 5 to 3. A third quite rational analysis might be that since the probability the second man goes on to win is only 1/8 (or $(1/2)^3$ by flipping three consecutive tails, he should receive only 1/8 of the pot and the first man 7/8 of it. Still other divisions are possible, all depending on the particular situation.

Clearly, mathematics can give us the consequences of whatever assumptions we make about the fairness of the split, and here those consequences are clear-cut and transparent. What mathematics can't do is adjudicate what is fair in any given situation.

This brings to mind some of the problems brought about by vastly more complicated and impenetrable calculations, specifically the widespread employment of machine learning and, more generally, artificial intelligence. In recent years, for example, Google has fed a humongous collection of sentences from a huge variety of sources—newspapers, books, the internet, articles, and so on—into a neural network, which is a set of algorithms intended to loosely reflect the way the human brain functions. The networks were primed to recognize patterns and correlations of all sorts.

Some of the patterns it observed and analogies it generated were troubling, however, especially since it was far from transparent exactly how they were generated. Asked to complete "man is to doctor as woman is to x," the x that was generated was "nurse." Likewise, "shopkeeper is to man as x is to woman," and x turned out to be "housewife." As Brian Christian observes in his book *The Alignment Problem*, there were more serious problems with such machine learning. One is that judges often have relied on risk assessment metrics that register only a few facts about individuals already in the criminal justice system and defendants who might soon be. Although these risk scores (the score was some whole number out of 10) have been widely used, researchers who carefully looked at them found that they often made no sense, were not predictive, had many false positives, and, as noted, were quite opaque.

The risk evaluations were sometimes racist, sexist, or simply not reflective of our better angels, baking in values and prejudices that we don't want or, at least, that we don't want to want. They were also one-size-fits-all and didn't allow for human oversight. It is this latter consideration that is most troubling because more and more of our decisions and actions are being determined by complicated algorithms to which we're blindly acceding. As useful as these and other uses of artificial intelligence often are, the lack of human oversight should alarm people.

In simple instances, the issue is a variant of the perennial problem of inappropriate models. There's the unfortunate story, for example, of a people who hunted deer with bows and arrows. Soon after mastering vector analysis, however, the technologically adept archers went hungry. Before learning about vector addition, when they spotted a deer to the northwest, they would shoot their arrow directly toward it. Afterward, when spotting a deer to the northwest, they would shoot one arrow to the north and one to the west, and the deer, of course, would scamper away.

A silly story perhaps, but it's silly only because we understand what's going on. When the silliness is masked by scores of unknown variables and parameters, huge data sets often of dubious quality, and self-generated algorithms, we just shrug and shoot one of our arrows to the north and one to the west.

The concern with machine learning and artificial intelligence generally is that we can't tell what is hardwired into the results that the systems yield. Will they suggest the analogue of telling us to shoot arrows to the north and west? What are their goals, limits, and purposes? Are they aligned with or ultimately antithetical to ours? And what, after all, are our goals, limits, purposes? We may recognize them as we go along, but we'll always revise them and always put them through a hard-to-specify human filter.

Without such a human filter and human oversight, employing superintelligent algorithms and systems might result in catastrophic consequences. Consider a well-known thought experiment devised by Nick Bostrom and referred to as the paperclip maximizer problem. If the goal built into an artificial intelligence system is to maximize paperclips, it will do so, of course, but it will also work to increase its own intelligence to enable it to produce more and more paperclips.

The system wouldn't value intelligence for its own sake but only as a means to optimize its ability to produce more paperclips.

Eventually, the system would go through an intelligence explosion that would result in the dwarfing of human intelligence. It would devise increasingly clever ways to maximize the number of paperclips, ultimately filling the whole planet and beyond with paperclips. Clearly, the future monomaniacal nature of such an algorithm innocently devised to maximize paperclips would not take into account human values, interests, and pursuits. It would be completely alien to us.

This is, of course, only a thought experiment, but like all thought experiments, it is intended to stimulate thought. In this case, the thought is of the future perils of artificial intelligence. We're all at risk of being bamboozled archers or being inundated by paperclips.

Proceeding from artificial intelligence to artificial intimacy, I ask why it is that prostitution across the world and back through history pays relatively quite well. Since the issue is quite topical, what sensible, non-preachy lessons can be inferred from the income of (many) sex workers?

SEXONOMICS: PROSTITUTES' INCOMES, A NONMORALISTIC ACCOUNT

The best-selling book *Freakonomics* examines the economics of some common life situations. If these situations involve sex, their analysis might be better termed "sexonomics." One of the first practitioners of sexonomics was Nobel Prize–winning economist Gary Becker, whose 1970s paper "A Theory of Marriage" pushed the economic analysis of sex into the purview of his fellow economists. More recently, two other economists, Lena Edlund of Columbia University and Evelyn Korn of Eberhard Karls University of Tübingen, have published an intriguing paper, "A Theory of Prostitution," in the *Journal of Political Economy*.

Making simplistic but more or less plausible assumptions and applying the tools of economic model making, they searched for the answer to a puzzle: Why is it that prostitution is so relatively well paid?

Before getting to why this is, they document that in diverse cultures and throughout many centuries, prostitutes have indeed made much more, sometimes several multiples more, than comparably (un)skilled women would make in more prosaic occupations. From medieval France and imperial Japan to present-day Los Angeles and Buddhist Thailand, this income differential is quite persistent, although its size depends on various factors.

(That prostitutes generally make considerably more money than their skills would warrant may be obscured by biased sampling. Studies of prostitutes often survey those in trouble with the law or on drugs and hence not earning what they might. By analogy, if studies of marriage were frequently conducted at shelters for battered wives, people would probably soon develop a more jaundiced view of marriage as well.)

Prostitution Diminishes Marriage Chances

Developing the consequences of their mathematical model, Edlund and Korn argue that the primary reason for the income differential is not the risk sometimes associated with the practice of prostitution but rather that prostitutes greatly diminish their chances for marriage by virtue of their occupation. Men generally don't want to marry (ex)prostitutes, and so women must be relatively well compensated to forgo the opportunity to marry.

Employing market concepts, doing some calculus, and assuming that "women sell and men buy," the authors also conclude that prostitution generally declines as men's incomes increase. Wives and prostitutes are competing "commodities" (in the reductionist view of economists, that is), but wives are distinctly superior in that they can produce children that are socially recognized as coming from the father. Thus, if men have more money, they tend to buy the superior "good" and, at least when wives and prostitutes come from the same pool of women, tend to buy (rent) the cheaper good less frequently.

More obvious perhaps is that prostitution generally declines in areas where women's incomes and opportunities are greater. Putting these two tendencies together suggests that if one wishes to reduce prostitution, increasing the incomes of both men and women is likely to be more effective than imposing legal penalties.

Sex Ratios, Foreign Prostitutes, and Cultural Factors

Another consequence of the authors' model is that a high ratio of men to women tends to increase prostitution's relative profitability (versus marriage). If the surplus of men over women is temporary, say, because of war or upheaval, then the surplus usually leads to an even greater incentive to prostitution. As permanent residents in a location, men are potential participants in both the marriage market and the sex market, whereas if they're visitors, only the latter market is generally available, and the supply of prostitutes and their incomes rise. The authors cite examples from 12th-century crusaders to modern sex tourists.

The model also predicts that how much a woman damages her chance to marry by becoming a prostitute depends on how likely it is that she'll be exposed as one. The likelihood shrinks if the woman leaves home and migrates to a different part of the country or to a different country altogether. This would also explain why foreign prostitutes are likely to be cheaper than domestic ones.

122

More generally, the abundance of foreign prostitutes shouldn't be a surprise. Immigrants generally have difficulty finding employment, and, except at the high end of the scale, prostitution does not place much of a premium on language skills. As in other parts of the economy, globalization is controversial and is one reason the number of women trafficked for sexual purposes is exaggerated. (It is considerably smaller than the number of people trafficked for nonsexual labor.) In fact, there are good reasons—from academic studies to the sheer ubiquity of prostitutes—to believe that this heinous practice is relatively isolated and that only a small fraction of prostitutes are coerced into prostitution.

One last prediction the model makes is that the income differential paid to prostitutes will rise with the status the culture accords wives. That is, if wives are valued highly, would-be prostitutes are giving up a lot by becoming prostitutes and will require more money to do so. And if wives have few privileges, would-be prostitutes aren't giving up much to become prostitutes and thus need less inducement to do so.

Cultural tolerance, of course, is a determinant not only of the income differential but also of the number of women who become prostitutes. Compare, for example, Thailand and Afghanistan.

Like any statistical model, this one ignores the diversity of real people and the complexities of love and pleasure, changing social mores, and so on. Still, once all its equations have been solved, a simple fact remains: Most women enter prostitution for the money. This being so, legalizing it, regulating it (strictly enforcing laws against trafficking, pimping, child prostitution, public nuisance, and so on), and improving the economic prospects for women seems to me to be a greatly preferable approach to it than moralistic denunciation.

Sex and lies, a planetary perspective.

SEX, LIES, AND STATISTICS: SOME MUSINGS
For individuals, sex is a private event whose frequency varies radically with circumstances: age, relationships, job stress, and so on. Looked at from a planetary perspective, however, the number of incidents of sexual intercourse initiated daily probably changes very little from one day to

the next, week in and week out, year after year. I would estimate the rate to be about 20 million episodes per hour with regular ups and downs due to differences in population densities in various time zones. As the great Polish science fiction author Stanislaw Lem mentions in *One Human Minute*, what's important is not the number, which is perhaps nothing more than a bad guess, but the amazing unrelentingness of the pattern whatever it is.

Sex surveys are often less insightful than a good novel or such idle musing. Results like those in the recent *JAMA* study "Sexual Dysfunction in the U.S." (bravely released on Valentine's Day) are conducted with scientific care, but I suspect that even these are seriously flawed. One major reason for both the blandness and the likely inaccuracy of these surveys is (surprise!) that people don't tell the truth about their sexual practices, especially to strangers. In most surveys, the average number of sex partners reported by heterosexual males, for example, is significantly greater than the average number reported by heterosexual females, and until the introduction of Viagra, the number of American men reporting erectile dysfunction appeared to be somewhere around 29. Consider too the MacArthur Foundation MIDUS study "Midlife Development in the U.S." on the seeming contentment of people in their middle years. (Question 1: Why might this study be systematically biased?)

But there is a partial remedy for this "problem" of lying. It is mathematically possible to obtain sensitive information about a group of people without compromising any person's privacy. If we want to discover what percentage of a large group of people have ever "X'ed"—betrayed a spouse, engaged in a certain sex act, or whatever—we can use the following technique, which has many refinements. Have every member of the group you wish to survey privately flip a coin. If the coin lands tails, the person is told to answer the question honestly: Has he or she ever X'ed: yes or no? If the coin lands heads, the person is directed to answer yes regardless of the correct answer to the question. Thus, a yes response could mean one of two things, one quite innocuous (obtaining a head), the other potentially embarrassing (X'ing). Since the experimenter can't know what the yes means, people presumably will be more honest in their responses. Mathematically, it's not difficult to infer from the percentage of yes responses what percentage of the population have X'ed without learning anything about any particular individual. (Question 2: If 580 out of 1,000 people answer yes, what percentage of this group have ever X'ed?)

Perhaps it's because I'm a mathematician that I sometimes prefer a priori propositions like Lem's to empirical surveys. One of my favorites involves the logical relation that connects two people, say, George and Martha, if there is a set of intermediate people such that George has had sex with A, who has had sex with B, who has had sex with C, and so on until Martha is reached. If we place all people who are related to one another in this way into the same group, this relation divides all Americans into nonoverlapping groups of people. My guess is that there are many celibates who are in their own single-person groups; a large number of monogamous couples neither of whose members ever had sex with anyone else and thus constitute their own two-person groups; many groups having three, four, five, or a relatively small number of members; and then the rest of the U.S. adult population in one huge group containing scores of millions of members. The vast majority of the latter group is not promiscuous, however. In fact, many of its members have had only one or a few partners. The size of the group derives from our interconnectedness.

Whatever the truth of these and other theoretical conjectures and factual inquiries, the only mathematical generalization about sex that has ever seemed totally credible to me is the one asserting that the average adult in the United States has one testicle and one breast. Whatever the meanings of the words "sex" and "is," sex is difficult to quantify.

Answer to question 1: My suspicion is that, despite the researchers' best efforts, the results of the MacArthur Foundation study on midlife satisfaction with life are skewed. The randomly selected respondents were questioned by strangers over the telephone and then were mailed a detailed follow-up questionnaire. Self-protective and jocular responses like "Oh, we're happy and everything's fine" are much more likely to be elicited by such an approach than are responses like "My job is a meaningless joke, and my spouse is from a different emotional planet."

Answer to question 2: Of the 580 people who answered yes to the question about ever having X'ed, approximately 500 answered yes because their coin came up heads. Thus, 80 of the approximately 500 people whose coins landed tails were truly answering the question in the affirmative. Without compromising any individual's privacy, we can infer that 80/500, or 16%, of the group whose coins landed tails have X'ed. Since there's no difference between those whose coins landed tails and those whose coins landed heads, this percentage can be taken to hold for the whole group of 1000 as well.

This argument by reductio ad absurdum shows that, were some biological facts to change, an adamant opposition to abortion would have to change as well. The argument undermines absolutist positions against abortion, although the proponents of such positions will probably not be swayed. The reference to the Supreme Court in the first sentence is still quite germane.

REDUCTIO: ABORTION THROUGH THE LOOKING GLASS

Although abortion battles are in the news with the nominations of new Supreme Court justices in recent months, all the arguments we hear about the issue are rather familiar and stale. In an effort to introduce a new albeit somewhat fanciful argument, let me begin with a classic story that is usually attributed to George Bernard Shaw.

Seated at a posh dinner party, Shaw asks the woman sitting next to him if she'd sleep with him for $1 million. She laughs and says she would, after which he asks her if she'd do so for $10. Outraged, she says, "What do you think I am?" He replies, "That has just been established. Now we're just haggling about the price."

Such hyperbolic extrapolations and exaggerations are useful when questioning the absoluteness of people's beliefs and so might be helpful with an issue like abortion, in which people often adopt an inflexible and dogmatic pro-life position.

Anti-abortion groups sometimes employ this technique in their skirmishes with pro-choice groups: If an abortion at two months is okay, they ask, why not one at six months? And if one at six months is acceptable, why not kill infants, toddlers, or the very old?

A more recent example once occurred during William Bennett's radio show. The former Reagan administration secretary of education got himself in trouble when a caller to his show prompted Bennett to refer to an intriguing and quite plausible argument made by the economist Steven Levitt. Levitt, the author of *Freakonomics*, maintained that the decline in crime in the 1990s was in large part the result of the significant increase in abortions in the 1970s.

Bennett found this hypothesis morally repellent. Wanting to show that even if the hypothesis were true it would still not justify abortion in his eyes, Bennett exaggerated the argument and introduced the element of race with the stated intent of showing the argument's deficiencies.

Not surprisingly, he was attacked as a racist. I'm not a fan of his political conservatism or of his moral unctuousness, but I don't see his argument as evidence of racism. He used a common sort of logical stratagem, which can easily be taken the wrong way by people not accustomed to it. Had he not been ad-libbing, he could easily have made a similar point without causing offense by bringing up the extraneous element of race.

In any case, at the risk of suffering similar attacks from a different swath of the political spectrum, consider the following argument, which also depends on a contrary-to-fact exaggeration to make its point. It's an argument that pro-choice proponents might use to undermine the belief of some abortion opponents in the absolute inviolability of the fetus's right to life.

We might ask them what position they would take were two biological facts to change. The first is that for some reason—a worldwide pandemic and an invincible new virus, a climatic catastrophe in the atmosphere, the effect of some new substance in the environment, whatever—pregnant women throughout the world would find themselves carrying 10 to 20 fetuses at a time. The second, more minor change is that advances in medical, surgical, and neonatal technology would make it possible to easily save some or all of these fetuses if doctors intervened a few months after conception. Without this intervention, however, all the fetuses would die.

We would have to assume that abortion opponents who believe that all fetuses have an absolute right to life would surely opt for some intervention. If not, all the fetuses would die. They would therefore have to choose to either adhere to their absolutist position and insist that all the fetuses live and thus be overwhelmed by a population explosion of overwhelming magnitude or else act to save only one or a few of the fetuses. But since all the fetuses are viable, the latter choice would be tantamount to abortion. It would take someone quite inflexibly dogmatic to opt to have the birthrate increase by a factor of 10 to 20.

Perhaps a fatuous argument, but the simple point is that if certain contingent biological facts were to change, then presumably even ardent abortion opponents would change their position, suggesting that their position is itself contingent and not absolute. After this is acknowledged, the haggling about the details might proceed.

More on the ubiquity of mismeasurement, this column concerns primarily Dan Brown's *The Da Vinci Code* and the movie of the same name. The novel is based on the premise that Jesus married and had children and that a descendant of his is alive today. The appeal of Brown's book depends critically on not making a distinction between genetic ancestry and genealogical ancestry. The latter refers to all of your ancestors no matter how distant or tenuously related (say, a 39th cousin five times removed), whereas the former refers only to the subset of them directly connected to you through your parents, grandparents, great grandparents, and so on. Incidentally, former vice president Dick Cheney and Barack Obama are ninth cousins once removed.

Also included in this column is a note about the preposterously high estimates of trafficked victims in the United States. One recent reason for the persistence of the belief that predators and traffickers are ubiquitous (including a good fraction of the Democratic Party) is due to QAnon, which has no doubt noticed that such a message and "estimate" are quite potent in appealing to prospective women members. Save the Children is a wonderful charity, but the phrase "save the children" is being used by other less selfless organizations to recruit members. Baseball is mentioned at the end to lighten the discussion.

MISUNDERSTANDINGS: JESUS' GENEALOGICAL DESCENDANTS, SEXUAL PREDATORS, AND HOME RUN RECORDS

Probability theory tells us that if Jesus had any children, his biological line would almost certainly have either died out after relatively few generations or else grown exponentially so that many millions of people alive today would be genealogical descendants of Jesus.

Of course, this is not a special trait of Jesus' descendants. If Julius Caesar's children and their descendants had not died out, then many millions of people alive today could claim themselves to be genealogical descendants of Caesar. The same can be said of the evil Caligula and of countless anonymous people living 2,000 or more years ago.

The research behind these conclusions, growing out of a subdiscipline of probability theory known as branching theory, is part of the work of Joseph Chang, a Yale statistician, and Steve Olson, author of *Mapping Human History: Genes, Race, and Our Common Origins*. (Much work on these matters is ongoing and sure to change our ideas significantly.)

Going back another millennium, we can state something even more astonishing. If anyone alive in 1000 BC has any present-day genealogical ancestors, then we would almost all be among them. That is, we are descended from almost *all* of the Europeans, Asians, Africans, and others who lived 3,000 years ago and have genealogical descendants living today.

It's interesting to look at this prospectively. If you have children and if your biological line doesn't die out, then almost every human being on Earth 2,000 or 3,000 years from now will be your genealogical descendant (but, importantly, not your genetic descendant).

Getting back to *The Da Vinci Code*, we can conclude that if the heroine of the book were indeed a genealogical descendant of Jesus, then she would share that status with countless millions of other people as well. But the heroine would almost certainly not be a direct *genetic* descendant of Jesus, which underlines the distinction between genealogical descent and genetic descendant, but that's another story hinted at above—and a long one at that. The failure to distinguish genealogical from genetic descendants undermines the book's plot, but then probability was never much of a match for fiction or Hollywood.

Sexual Predators

Announcing Project Safe Childhood, Attorney General Alberto Gonzales cited a frightening figure: "It has been estimated that, at any given time, 50,000 predators are on the Internet prowling for children." The only problem with this statistic is that it seems to have been made up out of whole cloth.

The phrase "at any given time" may be Gonzales's own bit of hyperbole, but his office cited media outlets that have focused on pedophiles and used the 50,000 statistic. TV shows and internet sites in turn cited law enforcement agents for the figure, and now the attorney general cites the media.

Jason McClure, a writer on legal affairs, recalls that in the 1980s, 50,000 was the number of people killed annually by satanic cults as well as the number of children kidnapped annually by strangers. Both of these numbers later proved baseless and absurdly high but perhaps derived some of their initial appeal from the roundness of the figure and its middling nature, neither too small nor too large.

In any case, this is another instance of a common phenomenon: A number gains a certain currency when commentator A pulls it out of . . . the air and is then cited by B as the number's source, B is cited by C as the number's source, C is cited by D, and so on, until someone in the loop is cited by A, and few ever check to see if the number has any validity, and often, of course, it doesn't. Still, for a while at least, "everyone knows" it's true.

Less Heavy: Home Run Records and Prime Numbers

Thirty-two years ago, Hank Aaron hit his 715th home run and surpassed Babe Ruth's record of 714, a feat that Barry Bonds accomplished in 2007. Reams of newsprint have been devoted to various aspects of these record breakings, but not well known is that Aaron's breaking of the record stimulated a spate of mathematical papers about what have come to be known as Ruth–Aaron pairs.

If we decompose 714 into its prime number factors, we find that $714 = 2 \times 3 \times 7 \times 17$. Likewise, $715 = 5 \times 11 \times 13$. Adding the prime factors of 714 and 715, we find that they have the same sum. That is, $2 + 3 + 5$

+ 17 = 29, and 5 + 11 + 13 = 29. Consecutive numbers like 714 and 715, whose prime factors sum to the same number, have come to be called Ruth–Aaron pairs.

Interestingly, if we multiply 714 (which equals 2 × 3 × 7 × 17) by 715 (which equals 5 × 11 × 13), we find that 714 × 715 = 2 × 3 × 5 × 7 × 11 × 13 × 17. So we have the product of the two consecutive whole numbers, 714 and 715, equal to the product of the first seven prime numbers, 2 × 3 × 5 × 7 × 11 × 13 × 17.

In any case, as is their wont, mathematicians tried to find out how common these properties were among pairs of consecutive numbers and whether there were arbitrarily large such pairs. (No one knows yet.) Even Paul Erdos, the famous peripatetic mathematician, wrote about Ruth–Aaron pairs (or Ruth–Aaron–Bonds pairs) and proved a crucial theorem about them.

Having grown up in Milwaukee, where Aaron played for years on the then Milwaukee Braves, I can't help but hope that "Hammering Hank's" record of 755 home runs stands even though the latter number (755 = 5 × 151) seems to be rather mathematically undistinguished.

CHAPTER FIVE

Partisanship in Politics

*There are men running governments that shouldn't be allowed to play
with matches.*

—WILL ROGERS

THERE ARE, OF COURSE, MANY MATHEMATICAL ASPECTS OF POLITICS,
ranging from technical results such as Arrow's theorem on the impossi-
bility of designing a voting system that can't be gamed to well-known and
traditional figures in political philosophy. Even Thomas Hobbes and his
philosophy of trading freedom for safety and John Rawls and his theory
of justice as equality and fairness give rise to some interesting mathematical and social issues. Herein, however, the bonbons presented center
around the narrow issue of the polarization of American politics and the
virtual Grand Canyon that divides the political landscape.

Former president Trump, of course, plays a starring role with paradox, ignorance, greed, and the internet in supporting roles. Unfortunately,
some of the columns on lies and lying are relevant here as well. Unless
some common narrative ground and adherence to truth is assumed,
democracy is doomed. There is no basis on which to base it if we remain
epistemic enemies. This debasement of truth has been well explored in
many recent books and articles, so I will attempt here to approach it
obliquely.

This perhaps tiresome list of Trumpian outrages is related to more general phenomena and is an illustration of Brandolini's law of refutation also known as the bullshit asymmetry principle.

TRUMP: OUTRAGE FATIGUE, DENIAL-OF-SERVICE ATTACKS, AND BRANDOLINI'S LAW OF REFUTATION

I will begin in a highbrow manner with Aristotle's theory of politics and the differences between a constitutional government, oligarchy, and democracy and the various types of states the Greek philosopher catalogs. No, I won't. I will begin with a lowbrow discussion of an idiocracy, a country reasonably characterized as a schizostan, and just a few of its former leader's outrageous actions, behaviors, and words. I can't help myself but you can by skipping the next few paragraphs.

In no particular order: Mexicans as rapists, kowtowing to Putin, abandoning the Kurds, disrespecting John McCain's service, firing inspector generals, pulling out of the Paris climate accords, extorting the Ukrainians, claiming massive voter fraud, separating kids from their parents and placing them in cages, countless false claims like the size of his inaugural crowds, skepticism of vaccines, the smearing of Joe Scarborough, the active promoting of birtherism for years, not releasing his taxes, violating the emoluments clause, spending more than $100 million on his golf trips while making a big deal of contributing his salary, his incompetent and corrupt cabinet appointments (to the Environmental Protection Agency, the Department of Health and Human Services, and the Department of Energy), personal attacks on the Khan family, a Mexican judge, Meryl Streep, his trillion-dollar tax cut that benefited primarily the wealthy, his association with liars and toadies from Giuliani and Manafort to Stone and Barr, his massive obstruction of justice, his dismissal of science leading to a catastrophic failure in dealing with COVID-19, outright lies ranging from being a friend of Pavarotti to not knowing Stormy Daniels (more than 20,000 according to the *Washington Post*), steadily weakening our ties to the European Union while cozying up to dictators like Putin and MBS, his aggressively clearing nonviolent protesters to enable a charade of a photo op, and so on.

Even his technically true statements are seriously misleading. A typical instance involved the high-tech notion of adding: In early July, after tens of millions of applications for unemployment insurance had been filed, Trump chortled that the economy had added a whopping 4 million new jobs in June. True, but if a big part of your house has burned down and you've started rebuilding it, you don't describe your efforts as putting an addition on your house.

Underlying these and so many other outrages big and small are his cruel narcissism and ignorant refusal to listen to anyone except the commentators on Fox News. Stalin is reputed to have said that a single death is a tragedy, a million deaths is a statistic. Paraphrasing here, I note that a single mistake, insult, or consciously false statement by a politician is a serious offense; 20,000 or more of such is a statistic. Hillary's emails versus Trump's malevolent and extensive web of lies and corruption. All right, if you're very astute, you may suspect I have TDS, Trump Derangement Syndrome. I do.

A less obvious observation about the partial list above that is supplemented almost hourly by Trump's scrofulous tweets is that it is somewhat akin to a denial-of-service attack, which is used to overwhelm a website with so many requests and bits of information that the site can't respond and shuts down. It may be a bit of a stretch, but the analogy to outrage fatigue, our tendency to grow numb to Trump's statements and behavior, is suggestive. We're a bit like the websites that shut down when overwhelmed by too many bits of information. Our attitude is too often, ah, that's just Trump being Trump again rather than alarm that that's just Trump undermining American democracy.

Quite germane to all this is Brandolini's law of refutation, first formulated by Italian programmer Alberto Brandolini. Sometimes described as the bullshit asymmetry principle, it states, "The amount of energy needed to refute bullshit is an order of magnitude bigger than to produce it." This can be thought of, perhaps, as the second law of political thermodynamics.

An earlier formulation by the Russian physicist Sergey Lopatnikov makes a similar point: "If the text of each individual phrase requires a paragraph (to disprove), each paragraph—a section, each section—a chapter, and each chapter—a book, the whole text becomes effectively irrefutable and, therefore, acquires features of truthfulness." The last clause seems particularly apt when applied to the truthiness of Trump, Facebook, and Fox News. Every bit of nonsense and disinformation about COVID-19 takes epidemiologists weeks and months to shoot down.

More generally, any rapid-fire sequence of lies, half-truths, misleading statements, ad hominem attacks, bad analogies, and irrelevant anecdotes will in the short run win out over a slow and painstaking fact check.

Trump, of course, had no shortage of others who regularly took advantage of Brandolini's law: lying enablers ensconced in a supportive subculture in which the disaffected, resentful, poorly informed, and greedy and self-serving special interests as well as the delusions of the fiercely independent everyman hold sway. Where else would so many people give seemingly equal weight to the statements of both entertainers and biological researchers on vaccines? Why are the pronouncements of both industry lobbyists and climate scientists accorded the same respect? And why are economic prescriptions, such as an unwarranted tax cut for the 1%, balanced against those from Nobel Prize–winning economists? Such an environment has proved a fertile ground for Trump's bloviating and malevolent incompetence.

And perhaps most of all is the unwavering unity of cowardly Republicans in the face of all his outrages. They should know better and are much more deserving of scorn and condemnation than the millions who just took Trump and his enablers at their word.

The account below is rather partisan, but I recognize that all political parties are at times subject to mindless partisanship and lockstep behavior. Here, however, I want to focus on the extreme case of the present-day Republican Party.

WOLF'S DILEMMA AND EXTREME LOCKSTEP POLITICAL PARTISANSHIP

Despite the many cruel, stupid, and destructive outrages, the Republicans in the House, in the Senate, and in the administration have been exceptionally united in their support of President Trump. They bring to mind a remark by Mark Twain: "Suppose you were an idiot and suppose you were a member of Congress; but I repeat myself." Suffering perhaps from professional myopia, I've asked myself if mathematics has anything to say regarding this partisan unity.

One answer, among many, is that political parties, especially in this contentious and polarized political environment, can face an interesting variant of the well-known prisoner's dilemma called Wolf's dilemma. Its gist is easy to present. A beneficent billionaire makes the following offer to a group of people. Each member of the group is told that they can secretly press the send key on their keyboards within 30 minutes or decide not to do so. If they all refrain from pressing this key, they will each receive a big award, say, $100,000, wired to their bank accounts. If even one person in the group presses the key, however, those who do will receive a smaller award, say, $10,000, while those who don't press the key will receive a pittance or even nothing.

Clearly, the members do not want to lose out on $100,000, which gives them a strong incentive to hang together and to exert psychological coercion on any would-be defectors to hang with them.

Political rewards are, of course, quite nebulous, whereas the dilemma's monetary awards are quantitative and well defined. Still, we can look at political instances of a somewhat similar predicament. Consider the Republicans in the House who pressured each other to maintain party unity in voting against an impeachment inquiry or the Republicans in the Senate who did the same by voting against a presidential conviction.

Wolf's dilemma suggests that pressing the key and defecting from the party in this political version would be equivalent to voting for an official impeachment inquiry in the House or voting to convict in the Senate and that refraining from pressing the key would be equivalent to voting against such an inquiry or against conviction.

The big reward for the unified Republicans would be a reduction in the chances of a presidential impeachment and removal and thereby increased chances of reelection and enhanced support of their constituencies. Ensuring the viability of Trump and maintaining their control of the presidency would be a boon to Republicans.

The smaller reward for those Republican House members who might have voted to begin the inquiry or to convict in the Senate would be an outpouring of praise for their courage from Democrats, Independents, and many pundits. Their colleagues would have a considerably gloomier political future because of this crack in their ranks, which throughout time might lead to a cascade of defections. And, of course, the beneficent billionaire could be replaced by a real not-so-beneficent billionaire or perhaps a super PAC.

It should also be noted that in the denatured Wolf's dilemma, one way to induce defections is to decrease the big award from $100,000 to, say, $50,000 and to increase the small award from $10,000 to, say, $30,000. Politics is a more visceral and unpredictable undertaking, but the analogues to this tactic in the political case should be clear: make sticking with Republican colleagues somehow less attractive and defecting from them more attractive.

No such tactic has worked so far. That the almost 200 Republicans in the House stayed unified is impressive since the larger the group and the weaker the bonds among its members, the more likely at least one of its members will vote to impeach the president. A similar dynamic prevailed in the trial in the Senate.

Alas, there is no easy "solution" to Wolf's dilemma. It is simply a possible logical skeleton of the situation the Republicans were in, which perhaps sharpens our focus on it a bit. Unfortunately, such lockstep party unity is not a rare phenomenon, and profiles in courage are.

Part of the reason for our political divisions is that our political districts at all levels are not rationally defensible.

Gerrymandering is unfortunately not generally acknowledged to be the scourge of democracy that it is. Although it's simply described as packing and cracking, its consequences are far reaching and not so simply described.

THROUGH A MATHNIFYING GLASS DARKLY

Gerrymandering is the partisan carving up of a state or other political entity into districts in a way that doesn't truly reflect the people in the state or other political entity. Such a neutral tone is at odds with the passionate partisanship the practice unleashes.

The essence of the practice and the related notions of packing and cracking is illustrated by the following simple numerical example. The party composition of the great state of Biastan is, let's assume, 53% Republican and 47% Democratic. Assume for the sake of simple arithmetic that the state has 1,000,000 residents and 5 congressional districts with 200,000 people in each district. If the Republicans are in charge of dividing the state into districts (as they are in most states), they might divide it thus:

District 1: 17D, 3R (170,000D, 30,000R), clearly a Democratic district

District 2: 8D, 12R (80,000D, 120,000R), a Republican district

District 3: 8D, 12R, the same composition, also Republican

District 4: 8D, 12R, the same composition, also Republican

District 5: 6D, 14R (60,000D, 140,000R), clearly a Republican district

The point of the redistricting by the Republicans should be obvious. Even though the state has a small Republican majority of 53%, 80% of the districts, 4 out of 5, will be safely Republican. Republicans have stacked the deck, but there is a complication. How do they achieve the above distribution geographically?

To get to the aforementioned redistricting, they likely will have to draw the districts in a very convoluted, snake-like way, including and excluding small areas and wandering in a seemingly crazy manner. They will be guided by two principles. One is that, where possible, they will pack as many Democrats as possible into a single district, such as in district 1 above. The second is that, where possible, they will crack Democratic areas into several different districts where they won't be a majority, as is the case in districts 2, 3, and 4 here.

By the judicious use of packing and cracking and creative boundary setting, Republicans and their elected officials (and the Democrats too in states in which they're in power) choose their voters as much as the voters choose them.

Gerrymandering is, in essence, a fancy word for rigging or cheating or manipulating, and it is manifestly unfair no matter which party does it. There are a number of mathematical constraints that would limit the effects of partisan gerrymandering. Among them are requirements that the districts be contiguous and compact and respect natural communities. Enforcing these and other restrictions, say, by a nonpartisan independent commission, has proved to be politically difficult to say the least.

The way in which states in the United States are accorded senators is, as mentioned, a kind of constitutionally mandated gerrymandering. States with very small populations, such as South Dakota and Wyoming, have the same representation, two senators, as states such as Texas and New York with very large populations. Shockingly, the 12 states with the smallest populations have a total combined population that is only one-third that of California, yet those 12 states are represented by 24 U.S. senators, California by two.

A feasible measure to at least partially counter this imbalance is to accord the District of Columbia state status. After all, it has more than 100,000 more residents than does the state of Wyoming.

A better way of voting that is more inclusive and open to a variety of candidates and less favorable to extreme candidates is ranked choice voting. Happily, it's being increasingly adopted in states and municipalities for elections where there are more than two candidates, such as primaries.

RANKED CHOICE VOTING: MORE WELCOMING TO MODERATE AND LESS CONDUCIVE TO EXTREME CANDIDATES
New York City is the latest political entity that has decided to use ranked voting to declare winners in elections with more than two candidates. How does it work? For illustration's sake, let's assume that there are five candidates for the position and there are only 55 voters (or 55 million)

and they rank their preferences (first, second third, fourth, and fifth choice) as follows:

18 members prefer A to D to E to C to B

12 members prefer B to E to D to C to A

10 members prefer C to B to E to D to A

9 members prefer D to C to E to B to A

4 members prefer E to B to D to C to A

2 members prefer E to C to D to B to A

Note there is no candidate supported by a majority of the 55 voters. However, if the standard plurality method is used to determine the winner, then candidate A is clearly the winner with 18 first-place votes, more than any of the other 4 candidates.

But which candidate wins if the ranked voting method is used?

Perhaps surprisingly, it's candidate C. We first eliminate the candidate with the fewest first-place votes. In this case, that's E, who has only 6 first-place votes. Then we redistribute the votes of E's voters, adjusting the first-place preferences for the others (still 18 for A, now 16 for B, now 12 for C, still 9 for D). In the next round, we repeat this and eliminate the candidate among the four remaining having the fewest first place votes. In this case, that's D. Then we redistribute D's voters, adjusting the first-place preferences for the others (still 18 for A, 16 for B, but now 21 first-place votes for C). We next eliminate the candidate with the fewest first-place votes, and that is B. Finally, we end up with a candidate, C, with the support of 37 of the 55 voters, a majority. And A, the winner if we used standard plurality method to determine the winner, ends up with his original 18 first-place votes.

In short, the voters rank their choices, and sequentially, we remove the candidate with the fewest first-place votes and redistribute that candidate's votes among the remaining candidates. We continue doing this until we get a candidate with a majority of the votes. Determining the winner takes a bit more time because with 5 candidates running, there are not just 5 possible ways the voters can rank them but 5! (or 120) ways.

Ranked choice voting gives voters whose clear preference isn't that popular a chance to nevertheless have some say in the election via their other choices. It also hurts those candidates who appeal to only a narrow subgroup of voters as candidate A above did. In general, it is better than the plurality method.

And Trump? Given that he was the plurality winner in the first Republican primaries with only about 30% of the vote, it is conceivable that ranked choice voting would have slowed the "Trump train" as it did the A train in the example above.

Unfortunately, there are still ways to use ranked choice voting to get ostensibly undemocratic results. In fact, there is a well-known mathematical result, Arrow's theorem, whose gist is that no voting method (and there are quite a few) satisfying certain minimal constraints is invulnerable. At times and in certain situations, they can all be scammed.

One relatively straightforward way to do so in the example above is to not vote your real preferences but to vote strategically instead. For the preference ranking above, that means that voters would lower their preferences for C and raise them for a candidate they find more congenial.

There is an analogy between the winner under plurality voting and the usual way of determining the hottest day of the year. To better define the latter or even just to get a better feel for how hot (or cold) a particular day was, we might take into account not just the highest temperature of the day but also the temperature for each hour during the day. In other words, we might imagine the ranked temperature preferences of the weather gods. So, for example, if one day the temperature was in the low 80s most of the day but with a high of 85, that day would be judged to be warmer than a day in which the temperature hovered around the mid-70s for most of the day but shot up to 88 for a couple of hours in the afternoon.

The messy consequences of the creaky electoral apparatus in Florida in 2000 predate Trumpian hyperpartisanship but can be considered a harbinger of them. Palm Beach County's vote totals for Pat Buchanan were a statistical "outlier," one that cost Al Gore the presidency. For want of a sensibly designed ballot, the election was lost. The difference between Gore's gracious concession and Trump's attempt to lead an insurrection couldn't be starker.

TIES AND COIN FLIPS: FROM BUTTERFLY BALLOTS TO BUTTERFLY EFFECTS

The Florida election is essentially a statistical tie. As I've written, the number of votes in dispute in this election is much greater than the difference in vote totals between the candidates. The voting apparatus is too gross to measure such tiny differences just as the scale we weigh ourselves on is useless for measuring micrograms of the vitamins we might take. Given this indeterminacy, the election might reasonably be settled by a coin flip.

Since such a decisive flip is unlikely to occur, we should ideally avoid the appearance of dredging for votes in partisan counties and hand count the votes in all the counties.

Still, given the closeness of the election and the margins of error and interpretations involved, any recount, even a careful manual one, of the entire state would be more or less tantamount to flipping a coin anyway.

To illustrate the iffy nature of the outcome, let me examine—after a short statistical detour—one of the continuing points of contention in the post-election campaign: the Buchanan vote in Palm Beach County.

There is a standard approach that statisticians use to understand the relationship between two variables. Take, for example, the heights and weights of people in one group or another. For each person in the group, one plots a point on a graph indicating his or her weight (say, on the vertical axis) and height (on the horizontal axis).

Using mathematical techniques that go by the name of regression analysis, one can find and draw the best-fitting straight line through these points. As common sense suggests, we would note that there is a positive relationship between the weight and height of people: The taller someone is, the heavier he or she generally is. There will, of course, be some "outliers"—very tall, light people or short, heavy ones—but these exceptions are unlikely to be extreme.

How is this relevant to the election? Since the vote totals for the candidates in each of the state's 67 counties are readily available, we can examine the relationship between the number of votes Reform Party candidate Patrick Buchanan received in a county and the number that Governor George W. Bush received in that county by following the same procedure.

For each county in Florida, we plot a point on a graph indicating the Buchanan vote (on the vertical axis) and the Bush vote (on the horizontal axis).

Applying the tools of regression analysis, we find and draw a line of best fit through the data and note that there is exactly one extreme outlier: Palm Beach County. It is so far away from the general drift of the data that it's somewhat analogous to finding a 700-pound person who is 5 feet, 6 inches tall in a group of 67 people.

We can also find the regression lines for the Buchanan vote versus the Gore vote or for the Buchanan vote versus the total vote, and again we would find that Palm Beach is the only extreme outlier. If our assumptions are correct and Buchanan's vote in the other 66 Florida counties is any guide, his vote total in Palm Beach is statistically quite extraordinary.

Furthermore, we can estimate with confidence that his total there was approximately 2,000 to 3,000 votes more than it should have been and deprived Gore of enough votes to throw the election to Bush.

The reason for the excessive Buchanan totals is no doubt the confusing "butterfly" ballot. Amusingly, it gives us a new illustration of the appropriately termed "butterfly effect" in chaos theory.

As mentioned, the term refers to the way a butterfly flapping its wings in Central America can lead to a snowstorm in New York a few months later. More generally, any small change occurring in some quantity can cascade toward a hugely disproportionate consequence down the road. In this case, the consequence is the identity of the next president of the United States.

The red states favored Bush, and blue states favored Gore. There are many reasons for this, but the simple model presented here suggests another one. People are conformist, and all politics is local, which suggests, incidentally, that political issues and disputes have a fractal structure.

VOTING BLOCS: RED STATES, BLUE STATES, AND A MODEL FOR THOUGHTLESS VOTING

Imagine a baseball reporter charged with covering the sport for his weekly newspaper. How long would he last if he gave the total number of runs produced by each team in the league during the week but seldom gave the number of games won and lost by each team?

Now imagine a political pollster charged with providing weekly updates on the electoral prospects of the candidates. How long would he last if he gave the percentage of voters nationally favoring each of the candidates but seldom gave the percentages in the important battleground states? It seems to me that the questions are quite analogous, but the baseball reporter would be viewed as a joke while the political pollster wouldn't be. Why? Especially in a close race, who wins the World Series of politics depends crucially on who wins the games in the individual purple states—those that are neither blue (Democratic) nor red (Republican) but somewhere in between.

But this is only marginally related to my topic. The question I want to consider is how there have come to be large contiguous regions of the country that are red or blue and only relatively small regions that are purple. Some light may be shed on this question by an abstract model introduced by Joshua Epstein of the Brookings Institution ("Learning to Be Thoughtless: Social Norms and Individual Computation"). I should mention here a caution provided by British statistician George E. P. Box: "All models are wrong, but some are useful."

Imagine that arrayed around a big circle are millions of people who are asked each day whether they intend to vote for George Bush or John Kerry. Assume that these people have an initial favorite, randomly choosing Bush or Kerry, but that they are very conformist and decide daily to consult some of their immediate neighbors. After polling the people on either side of them, they adjust their vote to conform with that of the majority of their neighbors.

How many people each voter consults varies from day to day and is determined by the fact that they are "lazy statisticians." They expand their samples of adjacent voters only as much as necessary and reduce them as much as possible, wishing always to conform with minimum exertion. There are various ways to model this general idea, but let's assume the following specific rule (which can be made more realistic). If one day a voter, say, Henry, polls the X people on either side of him, the next day, he expands his sample to the X + 1 people on either side of him. If the percentage favoring the two candidates in this expanded sample is different than it is when he polls only the X people on either side of him, he expands his sample still further.

On the other hand, if the percentage favoring the two candidates is the same in the expanded sample as it is when he polls only the X people on either side of him, Henry decides that he might be working too hard. In this case, he reduces his sample to the X - 1 people on either side of him. If the percentage favoring the candidates is the same in this smaller sample, he reduces the sample still further.

Every voter updates his or her favorite daily and interacts with other voters according to these same rules.

Epstein's model showed that the result of all this consulting is a little surprising. After several days of this sequential updating of votes, there are long arcs of solid Bush voters and long arcs of solid Kerry voters, and between these, there are small arcs of very mixed voters. After a short while, voters in the solid arcs need to consult only their immediate neighbors to decide how to vote and almost never change their votes. Voters between the solid arcs need to consult many people on either side of them and change their vote quite frequently.

Although Epstein didn't directly apply his model to voting but rather to more automatically followed social norms, the idea of extending it to voting is seductive. People do tend to surround themselves with others of like mind, and generally only those at the borders between partisans, the so-called swing voters, are open to much change. His major point, which I'm distorting a little here by casting his model into an electoral framework, is that social norms, often a result of nothing more than propinquity, make it unnecessary to think much—about what to wear, which side of the road to drive on, when to eat, and so on.

To the considerable extent that voting is, at least for many, an unthinking emulation of those with whom they associate, the model helps explain the near uniformity of the political opinions of their friends. (Rush Limbaugh's depressingly telling phrase "ditto heads" applies to many on both sides of the political spectrum.)

When there's some sort of shock to the system, Epstein's model suggests something else rather interesting. If a large number of voters change their vote suddenly for some reason (say, a terrorist attack or environmental catastrophe), the changed voter preferences soon settle down to a new equilibrium just as stable with solid Bush, solid Kerry, and mixed border areas but located at different places around the circle. The model thus shows how political allegiances can sometimes change suddenly but then settle quickly into a new and different segmentation just as rigidly adhered to as the old.

Returning finally to my introductory point, I note that unless there is some cataclysmic change in the presidential race, the only polls that count are the ones in Ohio, Florida, and the relatively few other contested states.

The internet has supercharged various universal human foibles, including the counterintuitive and pervasive conjunction fallacy. They help explain the rise in "fake news."

THE INTERNET, CONSPIRACY THEORIES, AND COGNITIVE FOIBLES, INCLUDING THE CONJUNCTION FALLACY

There are many reasons for the rise of conspiracy theories in recent years. Some of them derive from cognitive foibles we're all subject to, a couple of them well known, one rather less so.

One of these foibles is the anchoring effect, the tendency we have to be anchored to the first numbers we hear concerning any phenomenon. Trump, for example, said a number of times that were it not for his visionary leadership, COVID-19 would have killed millions of people. Judging by this landmark, he probably hoped his inexcusable negligence that has resulted in more than half a million (as of this writing) dead Americans wouldn't seem so sickening.

The availability error is also relevant. It is our tendency to be unduly influenced not only by the numbers we hear but also by the easy availability of vivid images that color our perception: Mexican caravans, Muslims dancing on the roof after 9-11, immigrant murderers in sanctuary cities, and so on. Innumeracy, of course, is also a factor as demonstrated by the difference between the value of his much-touted donation of his salary to charity and the exorbitant cost to the taxpayer of the emoluments, profits from his business, and travel expenses.

And let's not forget confirmation bias, which is easier than ever to fall victim to given the torrential current of unfiltered blogs, tweets, postings, and so on that appear on the internet. Most people naturally search for those sites that bolster their beliefs and biases and tend to ignore those that run counter to them.

Finally, there is a less well-known cognitive foible, the so-called conjunction fallacy, that is relevant. Let me illustrate with an example.

George is Norwegian, a kind man who loves animals and lives in the woods not too far from Oslo. He is physically fit, sometimes writes software from home, is divorced, and has two young children living with his ex-wife. Which is more likely?

1. George makes money by telling friends how to evade taxes.

2. George makes money by telling friends how to evade taxes and takes long walks with his dog in the countryside.

The second option sounds more likely because it sounds like a more coherent story being fleshed out. Despite that, however, the first option is more likely. The reason: For any statements A, B, and C, the probability that A is true is always greater than the probability that A, B, and C are all true since whenever A, B, and C are true, so is A but not vice versa.

More generally and more importantly, there is a trade-off between plausibility and probability: More plausible scenarios are often more detailed and hence less probable than simple unadorned descriptions.

What does this have to do with conspiracy theories? Given the plethora of factoids, speculations, rumors, hallucinations, distortions, exaggerations, and other dubious items on the internet, it is increasingly easy to construct elaborate scenarios that possess a degree of somewhat plausible detail. (After all, existential statements can't be conclusively disproved but might be proved, and universal statements can't be conclusively proved but might be disproved.)

Among countless possible outlandish narratives is the now almost canonical PizzaGate story. The rumor that Hillary Clinton and some of her top aides were involved in child trafficking and other crimes grew out of a good number of details that some partisans cobbled together because they were so wedded to their belief that Hillary and associates were atheistic sybarites or some variant of such nonsense. The details were either fabricated or taken out of context, a practice that QAnon has mastered and effectively weaponized.

The same can be said about the belief of some that Bill Gates is pushing vaccines to enable him and the government to inject a chip into everyone's bloodstream that would allow the government to collect people's information and thereby control them. Again, given bits and pieces from the internet, variants of this story have been cobbled together and fit the preconceived attitudes of many. Interestingly, few of the subscribers to this theory have noticed that Google already does that albeit via our eyes and not our blood.

Consider also the protests related to the killing of George Floyd. A wide range of narratives about it is possible. Does one emphasize the police brutality, the idealism of young people, and the increased nationwide awareness of institutional racism, or does one focus on the fires, looting, and violence of a few of the demonstrations. Are we aspiring to certain ideals or regressing to meanness? There are certainly a slew of dubious stories that stress one or the other of these choices that one might construct from factoids and commentary on the internet.

Finally, an interesting exercise: Pick a public figure that you despise or at least find distasteful. Make up a ridiculous story that would humiliate and embarrass him or her. Then look up items on the internet that can be selectively used to bolster the plausibility of your story. Artfully put them together, and, voila, you either have a new conspiracy theory or just an example of political opposition research.

The recent kudzu-like growth of conspiracy theories, "fake news," and QAnon nonsense mentioned above should not blind us to the fact that these human tendencies, although amped up and galvanized by the internet and modern media, have been around probably for as long as we have. Here is an example stimulated by the World Trade Center attack on 9-11.

AFTER 9-11: MINDLESSLY SEARCHING FOR NUMEROLOGICAL MEANING IN THE MIDST OF TRAGEDY

In times of crisis and heartbreak, many people's need for explanations of any sort seems to make them more open to the appeal of prophecies and coincidences. The Kennedy assassination, for example, led to a long list of seemingly eerie historical and numerical links between Kennedy and Lincoln. Accepting such sham explanations can be more comforting than facing the awful acts directly, puzzling out their causes, and framing our responses. This may be part of the reason for the outpouring of superstition that sprang up on the internet after the attack on the World Trade Center.

First there were the 11 numerologists whose emails began by pointing out that September 11 is written 9-11, the telephone code for emergencies. Moreover, the sum of the digits in 9-11 (9 + 1 + 1) is 11; September 11 is the 254th day of the year; the sum of 2, 5, and 4 is 11; and after September 11, there remain 111 days in the year. Stretching things even more, the emails noted that the twin towers of the World Trade Center look like the number 11, that the flight number of the first plane to hit the towers was 11, and that various significant phrases, including "New York City," "Afghanistan," and "The Pentagon," have 11 letters.

(Side note: The emails neglected to mention that 9-11 has a twinning property in the following rather strained sense: Take any three-digit number, multiply it by 91 and then by 11, and, lo and behold, the digits will always repeat themselves. Thus, 767 × 91 × 11 equals 767,767. Why? See below for the answer.)

There are many more of these after-the-fact manipulations, but the problem should be clear. With a little effort, we could do something similar with almost any date or any set of words and names.

The situation is analogous to the Bible codes. People search the Bible for equidistant letter sequences (ELSs) that spell out words that are relevant to an event and that can be said to have "predicted" it. (ELSs are letters in a text, each separated from the next by a fixed number of other letters.) Consider the word "generalization" for an easy example. It contains an equidistant letter sequence for "Nazi," as can be seen by capitalizing the letters in question: geNerAliZatIon.

There were emails and websites claiming the Bible contains many ELSs for "Saddam Hussein," "bin Laden," and also much longer ones describing the heinous acts at the World Trade Center. Unlike the original Bible codes, whose faults were rather subtle, these longer ELSs are purely bogus.

The most widely circulated of the recent email hoaxes involves the alleged prophecies of the 16th-century mystic and astrologer Nostradamus. Many verses were cited, most complete fabrications. Others were variations on existing verses whose flowery, vague language, like verbal Rorschach inkblots, allows for countless interpretations.

One of the most popular was this: "The big war will begin when the big city is burning on the 11th day of the 9th month when two metal birds would crash into two tall statues in the city and the world will end soon after." Seemingly prescient, this verse was simply made up, supermarket-tabloid style.

The truly ominous aspect of Nostradamus's prophecies was that it reached the number one spot on the Amazon best-seller list in the week after the attacks and that five other books about Nostradamus were in the top 25. Search engines were also taxed by surfers seeking out "Nostradamus," which temporarily even beat out "adult" and "sex" in popularity.

All of these hoaxes and coincidences involve seeing or projecting patterns onto numbers and words. Photographs brought out the same tendency in some who thought they saw the "devil" in the clouds above the World Trade Center or in the smoke coming out of it. These photos also appeared on many sites.

The reading of significance into pictures and numerical and literal symbols has a long history. Consider *I Ching* hexagrams, geometric symbols that permit an indefinitely large number of interpretations, none of which is ever shown to be correct or incorrect, accurate or inaccurate, predictive or not predictive.

Numerology, too, is a very old practice common to many ancient and medieval societies. It often involved the assignment of numerical values to letters and the tortured reading of significance into the numerical equality between various words and phrases. These numerical readings have been used by Greeks, Jews, Christians, and Muslims not only to provide confirmation of religious doctrine (666, for example) but also for prediction, dream interpretation, and amusement and as aids to memory and positive associations.

All people search for patterns and order, but some are determined to find them whether they're there or not. Sometimes it's hard to tell. If one flips a coin many times in succession, for example, and colors the successive squares of a large checkerboard black or white depending on whether the coin lands heads or tails, the resulting randomly colored checkerboard will frequently appear to contain a representation of some sort.

But human affairs are much more multifaceted than checkerboards. There are so many ways in which numbers, names, events, organizations, and we ourselves may be linked together that it's almost impossible that there not be all sorts of meaningless coincidences and nebulous predictions. This is especially so when one is inundated with so much decontextualized information (as on the internet) and overwhelmed by so much grief, fear, and anger.

The more difficult question is not why so many counterfeit connections were discovered but rather why some ominously real ones were not.

Answer to sidebar question: $91 \times 11 = 1{,}001$, which, when multiplied by any three-digit number, has the stated effect of $1{,}001(XYZ) = XYZ{,}XYZ$.

CHAPTER SIX

Religious Dogmatism

All you need in this life is ignorance and confidence, and then success is sure.

—MARK TWAIN

IN 2008, I WROTE *IRRELIGION*. IN THE BOOK, I EXPLORED WHY THE VARious arguments for God just don't add up. I tried to write a short, direct, and genial account that avoided the angry and denunciatory tone of many other books on atheism. The various standard arguments for the existence of God are briefly laid out and debunked, all of them shown to have fairly obvious flaws and gaping lacunae. Oddly, the most persuasive argument for many is the weakest: God exists because His assertion that He exists and discussion of His various exploits appear in this book about Him that believers say He inspired.

In any case, it's always been my belief that the heavy burden of proof for the extraordinary claims of religion should be borne by the claimants and that atheism or agnosticism should be the default stance.

I've included columns below that at least border on the above concerns. I don't mean to offend religious people, nor do I mean to dismiss or minimize the contributions of religion to art, music, literature, or, for some, as a way to escape the cages of their egos. In *Irreligion* and, to a limited extent, here, I simply try to present my views and the compelling reasons for them, compelling at least to me. As the poet Howard Nemerov said, "There's a long way from I Deny to

I Don't Know to Adonai," and there are good people in each of these nebulously defined groups. The problem is that, as Voltaire observed, "If we believe absurdities, we shall commit atrocities." This is not an idle warning today when a significant fraction of a major political party is the most potent promoter of absurdities on the planet. Religious dogmatism and extreme political partisanship are not unrelated. They're both manifestations of a swaggering ignorance and a mindless adherence to vacuous slogans.

I would hope that it's needless to observe that "Whatever God wills, happens" is an empty unfalsifiable assertion, but, as the philosopher Karl Popper frequently observed, unfalsifiability is rife not only in religion but also in everyday life and even in some "sciences." He cites as examples parts of Marxist economics and Freudian psychology. A Marxist might predict that the "ruling class" will respond in a grasping and greedy way to any economic crisis, but if it doesn't, he will attribute this to some self-regarding, co-opting policy of the ruling class. Similarly, a Freudian analyst might predict a certain category of neurotic behavior in response to a patient's personal crisis, but if the patient behaves in a quite contrary way, he will attribute this to some sort of "reaction-formation." The predictions, he concluded, are unfalsifiable. Popper isn't snarky, so he doesn't quite say so, but I suspect he thought that Freud was prone to flimflam, and Marx often made less sense than Groucho.

That the flawed notion of spontaneous order can be applied to economics as well as to biology undermines one of the basic contentions of "intelligent design" and "creationism." "Spontaneous order" is better termed "slowly evolving natural order." A different argument relying on simple probability also undercuts creationist claims.

NATURALLY EVOLVING ORDER AND PROBABILITY VERSUS THE CLAIMS OF INTELLIGENT DESIGN

The theory of intelligent design, which some tout as the more scientific descendant of creation science, rejects Darwin's theory of evolution as being unable to explain the complexity of life. How, ask supporters of intelligent design, can biological phenomena like the clotting of blood or the visual acuity of eyes have arisen just by chance?

The First Argument

One of the proponents of intelligent design likens what he terms the "irreducible complexity" of such phenomena to the irreducible complexity of a mousetrap. He argues that the mouse trap is a unitary entity and that should one of its pieces be lacking—the spring, the platform, or the metal bar—the trap is no longer a trap. The unstated suggestion is that all the parts of a mousetrap would have had to come into being at once, else there'd be no mousetrap, something that couldn't happen unless there were an intelligent human designer.

Intelligent design supporters maintain that something quite similar holds for more complex biological phenomena. If any of the many proteins involved in blood clotting were absent, clotting wouldn't occur, or if light-sensitive cells and any of their various necessary neural connections were missing, sight wouldn't occur. The creationist argument continues that since clotting and sight and countless other biological phenomena do occur, they had to have been created all at once by an intelligent designer.

This is a minor variant of William Paley's argument from design, which has been refuted many times. My intention here, however, is to develop some loose analogies between these biological issues and related economic ones and to show that these analogies point to a surprising crossing of political lines.

As I noted in *Irreligion*, it's natural to wonder why these creationist arguments don't apply to the complexity of free market economies. How is it that everything you might want to buy is available in stores nearby or else on Amazon? Your favorite brand of fig bars is minutes away. This amazing complexity is also apparent in industry. Despite occasional shortages, there are almost always enough steel girders, computer chips, and supplies of gas available for society to function more or less smoothly.

Likewise with communication networks and your e-mail, which reaches you in Manhattan as well as in Milwaukee, not to mention Buenos Aires and Bangkok.

The natural question, discussed first by Adam Smith and later by Friedrich Hayek and Karl Popper, among others, is, Who designed this marvel of complexity? Which commissar decreed the number of packets of Snickers for each retail outlet?

The answer, of course, is that there wasn't one. The system emerged and grew by itself, an example of—guess what—gradually yet spontaneously evolving order. What is more than a bit odd, however, is that some of the most ardent opponents of Darwinian evolution—for example, many fundamentalist Christians—are among the most ardent supporters of the free market. They accept and even champion laissez-faire and the natural complexity of the market yet continue to insist that the natural complexity of biological phenomena requires a designer and refuse to believe that natural selection and "blind processes" can lead to similar biological order arising spontaneously.

These ideas are not new. Adam Smith, Friedrich Hayek, Karl Popper, and others have made them more or less explicit. There are, of course, quite significant differences and dis-analogies between biological systems and economic ones (one being that biology is a much more substantive science than economics), but these shouldn't blind us to their similarities or mask the obvious analogies.

The Second Argument

Another simple argument, among so many that undermine creationist ideas, depends on a bit of probability, ignorance of which is just one contributing factor in the widespread American opposition to the theory of evolution. In an attempt to dress their opposition to evolution in mathematical garb, creationists point to the minuscule probability of evolutionary development. They argue that the likelihood of, say, a new species developing is absurdly tiny. The same, they say, is true of the development of the eye or the mechanism for blood clotting.

A standard creationist tack is to argue that very long sequences of improbable events must occur for a particular species or any biological phenomena to evolve. Moreover, if we assume these events are independent, then the probability that they all will occur is simply the product of these probabilities, which, of course, leads to a tiny probability.

Thus, for example, the probability of getting a five 12s in a row when rolling a pair of dice is $1/36 \times 1/36 \times 1/36 \times 1/36 \times 1/36$, or $(1/36)^5$, which equals .000000017—1 chance in 60,466,176. The much longer sequences of fortuitous events necessary for a new species or a new process to evolve lead to the minuscule numbers that creationists maintain prove that evolution is so wildly improbable as to be essentially impossible.

This line of argument, however, has a fatal flaw. Let's forget about whether the imagined events are independent; we should appreciate that there will always be a humongous number of evolutionary paths an organism or process might take. Yet there will be only one path actually taken.

So what? Well, we can focus on the path taken and ask ourselves what the probability is that this actual path was taken. The answer is the minuscule probability creationists like to tout when they reject evolution, but that's not the proper focus.

Another example illustrates this. Imagine we have a deck of cards before us. There are almost 10^{68}—a 1 with 68 zeros after it—orderings of the 52 cards in the deck. Any of the 52 cards might be first, any of the remaining 51 second, any of the remaining 50 third, and so on. This is also a humongous number, but it's not hard to devise even everyday situations that give rise to much larger numbers.

Now we shuffle this deck of cards for a long time and then examine the particular ordering of the cards that happens to result. We conclude, of course, that the probability of this particular ordering of the cards having occurred is approximately 1 chance in 10^{68}. This certainly qualifies as minuscule. Still, it would be silly for us to conclude that the shuffles could not have possibly resulted in this particular ordering because its minuscule a priori probability is so very tiny. Some ordering of the cards had to result from the shuffling, and this one did. There is no justification for the conclusion that the whole process of moving from one ordering to another via shuffles (that is, of evolving) is so wildly improbable as to be practically impossible, especially since the shuffles, to continue with the metaphor, generally involve only a relatively few cards (or genes).

The actual result of the shuffles will always have a minuscule probability of occurring, but, unless you're a creationist, that doesn't mean the countlessly possible results obtained via shuffling are at all dubious.

This latter is a special case of a common phenomenon: The probability of a particular outcome is low to quite tiny, while the probability of a general outcome of a somewhat similar type is high to almost sure. Unfortunately, people regularly confuse the two.

From "biblical mathematics" to Eugene Wigner, the two parts of this column may seem jarringly dissimilar, but both originate from a somewhat similar quasi apotheosis of mathematics.

ON THE QUASI APOTHEOSIS OF MATHEMATICAL IDEAS, TWO ACCOUNTS—ONE SILLY, THE OTHER SERIOUS

First, some nonsense about math education in certain benighted schools and then the profound question about the origin and effectiveness of mathematical ideas.

School begins again, and we read more about the intrusion of pseudoscience into school science curricula in this country, particularly into the study of biology and evolution.

The motive, despite the claims of proponents of intelligent design and other bogus "disciplines," has been religious. Recently, a number of readers have sent me course descriptions from various schools that bear this out.

Consider first a Baptist school in Texas whose description of a geometry course begins,

> Students will examine the nature of God as they progress in their understanding of mathematics. Students will understand the absolute consistency of mathematical principles and know that God was the inventor of that consistency. They will see God's nature revealed in the order and precision they review foundational concepts while being able to demonstrate geometric thinking and spatial reasoning. The study of the basics of geometry through making and testing conjectures regarding mathematical and real-world patterns will allow the students to understand the absolute consistency of God as seen in the geometric principles he created.

I wonder if the school teaches that non-Euclidean geometry is the work of the devil or at least of non-Christians. The website's account goes on like this for a while and then is followed by similar descriptions for algebra and precalculus. The blurb for the calculus course states,

> Students will examine the nature of God as they progress in their understanding of mathematics. Students will understand the absolute consistency of mathematical principles and know that God was the inventor of that consistency. Mathematical study will result in a greater appreciation of God and His works in creation. The students will understand the basic ideas of both differential and integral calculus and its importance and historical applications. The students will recognize that God created our minds to be able to see that the universe can be calculated by mental methods.

I don't know what books this particular school uses, but I should mention such risible texts as *Precalculus for Christian Schools*. The latter attempts to draw parallels between the fundamental theorem of calculus and the fundamentals of Christianity, between infinity and life after death, and so on.

Everyone's heard of church schools and madrassas, but another example of this phenomenon from a quite different religious perspective is the Maharishi University in Iowa, whose course titles and descriptions are similarly bizarre. Here are some on their website:

Infinity: From the Empty Set to the Boundless Universe of All Sets

Exploring the Full Range of Mathematics and Seeing Its Source in Your Self

Intermediate Algebra: Using Variables to Manage the Total Possibility of Numbers and Solve Practical Problems

Its New Age calculus sequence is described thus:

Calculus 1: Derivatives as the Mathematics of Transcending, Used to Handle Changing Quantities Calculus

2: Integrals as the Mathematics of Unification, Used to Handle Wholeness Calculus

3: Unified Management of Change in All Possible Directions Calculus

4: Locating Silence within Dynamism

From Fatuous to Profound, Switching Gears

Let me switch gears and note that there are, of course, much more sophisticated ideas that are at least vaguely similar. Moreover, there have been first-rate scientists who have taken mathematics to be some sort of divine manifestation. One of the most well known such arguments is due to physicist Eugene Wigner. In his famous 1960 paper "The Unreasonable Effectiveness of Mathematics in the Natural Sciences," he maintained that the ability of mathematics to describe and predict the physical world is no accident but rather is evidence of a deep and mysterious harmony.

But is the usefulness of mathematics really so mysterious? There is a quite compelling alternative explanation why mathematics is so useful. We count, we measure, we employ basic logic, and these activities are stimulated by ubiquitous aspects of the physical world. The size of a collection (of stones, grapes, animals), for example, is associated with the size of a number, and keeping track of it leads to counting. Putting collections together is associated with adding numbers and so on. Of course, the universe must be such that it contains identifiable and persistent "things" of one sort or another.

Another metaphor associates the familiar realm of measuring sticks (small branches, say, or pieces of string) with the more abstract one of geometry. The length of a stick is associated with the size of a number (once some segment is associated with the number 1), and relations between the numbers associated with a triangle, say, are noted. (Scores of such metaphors underlying more advanced mathematical disciplines have been developed by linguist George Lakoff and psychologist Rafael Nunez in their book *Where Mathematics Comes From*.)

Once part of human practice, these various notions are abstracted, idealized, and formalized to create basic mathematics, and the deductive nature of mathematics then makes this formalization useful in realms to which it is only indirectly related.

The universe acts on us, we adapt to it, and the notions that we develop as a result, including the mathematical ones, are in a sense taught us by the universe. That great bugbear of creationists, evolution has selected those of our ancestors (both human and not) whose behavior and thought are consistent with the workings of the universe. The usefulness of mathematics is thus not so unreasonable.

There are, of course, many other views of mathematics (Platonism, formalism, and so on), but whatever one's philosophy of the subject, the curricula cited above and others like them are a bit absurd, even funny. In private schools, they're none of our business. This is not so if aspects of these "creation math" curricula slip into the public schools, a prospect no doubt devoutly wished for by some.

In Genesis, there are "hidden references" to Roswell, New Mexico, and UFOs, but that doesn't mean anything. There are also probably hidden references to Roswell and UFOs in *Huckleberry Finn* as well.

An unfortunate corollary of the biblical injunction "seek, and ye shall find" is "make up stuff and get the conclusion you want." More on the Bible codes.

Holy Cow or Bull? ELSes: It Would Be Astonishing If One Didn't Find Hidden Messages in the Bible

Anxiety rises as we approach the nonevent scheduled for January 1, 2000. In an effort to lessen this millennial angst, I would like to update the situation regarding the Bible codes.

If you recall, the Bible codes are seemingly significant words spelled out by picking out every nth letter from a passage in the Bible. These equidistant letter sequences, ELSes for short, are alleged to foretell events that occurred long after the books were written. There are Christian and Jewish variants as well as Islamic analogues in the Koran.

Bible Predicts UFO Crash!

Here's an amusing example discovered by David Thomas, a physicist who has studied the frequency of these equidistant letter sequences.

Look at this passage from the King James version of Genesis:

> And hast not suffered me to kiss my sons and my daughters?
> Thou hast now done foolishly in so doing.

Start with the "r" in "daughters" and pick out every fourth letter: And hast not suffered me to kiss my sons and my daughteRs? ThOu haSt noW donE fooLishLy. In so doing, you have R-O-S-W-E-L-L.

Then start with the "u" in "thou" and pick out every 12th letter: And hast not suffered me to kiss my sons and my daughters? ThoU hast now done Foolishly in sO doing. You have U-F-O.

Could the Bible really contain a coded prediction of the supposed UFO crash in Roswell in 1947? Chances are, absolutely not, to say the least.

Equidistant letter sequences have attracted an inordinate amount of attention since *Statistical Science*, a leading international journal, published in 1994 an article titled "Equidistant Letter Sequences in the Book of Genesis." The article purported to find links in the Torah between the names of rabbis and prophets, specific dates, and historical events.

The mathematically sophisticated piece by Doron Witzum, Eliyahu Rips, and Yoav Rosenberg described many sequences in the Hebrew text of Genesis and calculated the probability of their occurrence to be vanishingly small, leading some to assert the codes' divine origin.

The editors of the journal published their results as a "challenging puzzle" since the explanation for the superficially remarkable ELSes was unclear. With this less than ringing scholarly imprimatur, the findings were soon elaborated on in author Michael Drosnin's 1997 best-selling book *The Bible Code* and in countless popular accounts.

Although most religious people and virtually all nonreligious people dismissed them, the codes convinced more than a few to become devoutly observant.

Refuting the Findings, Assuming They Need Formal Refutation

Because of the usual media practice of sensational headline allegations followed much later by barely noticeable corrections, you probably haven't heard of the detailed, 45-page refutation that was published in September. The new paper, "Solving the Bible Code Puzzle," was written by Brendan McKay of the Australian National University and Dror Bar-Natan, Maya Bar-Hillel, and Gil Kalai of Jerusalem's Hebrew University, and it appeared in the same journal, *Statistical Science*.

Here is part of the paper's abstract: "In reply, we argue that Witztum, Rips and Rosenberg's case is fatally defective, indeed that their result merely reflects on the choices made in designing their experiment and collecting the data for it. We present extensive evidence in support of that conclusion. We also report on many new experiments of our own, all of which failed to detect the alleged phenomenon."

The journal editors add that McKay and his associates found that "the specifications of the search (for hidden words) were, in fact, inadequately specific. . . . Because minor variations in data definitions and the procedure used by Witztum et al produce much less striking results, there is good reason to think that the particular forms of words those authors chose effectively 'tuned' their method to their data, thus invalidating their statistical test."

Timely Example

If, to use an example I've written about elsewhere, you were to look for prophetic evidence of the Clinton sex scandals in the Constitution, you could look for ELSes that begin anywhere within the document, that have any number of skips between letters of the ELSes, or that involve any words spelled backward, diagonally, or any which way. With so much leeway, it would not be too unlikely if you found some ELSes that spell out, say, M-O-N-I-C-A or P-A-U-L-A or G-E-N-N-I-F-E-R or K-A-T-H-L-E-E-N, or something similar—seemingly remarkable yet utterly insignificant.

The moral of the story: If you look long enough and hard enough, you'll likely find what you're looking for. Moreover, if you don't set up strict rules beforehand for searching the data and if you throw away all of the boring nonresults, then the interesting sequences that pop up by chance do not mean what they seem to.

As is even clearer now, the sequences that constitute the Bible codes mean nothing.

The McKay, Bar-Natan, Bar-Hillel, and Kalai paper provides a fitting coda to the codes phenomenon. Too bad all millennial worries can't be so easily disposed of. (Incidentally, there is an ELS for Y2K in the first two sentences of this piece.)

Solution:

Would an ELS with big skips between letters or one with short skips be more likely to be destroyed by small corruptions of the text? The authors of the paper refuting the Bible codes observe that adding or deleting a single letter in the text within the span of an equidistant letter sequence would likely destroy it. Thus, ELSes with big skips between letters would more likely be destroyed by minor variations than would ELSes with short skips.

This is not the case for the Bible codes, many of which have very large skips. (In *The Bible Code*, Michael Drosnin's ELS for the Hebrew equivalent of Y-I-T-Z-H-A-K R-A-B-I-N skips 4,771 letters between significant letters.) That undermines the argument of some that corruption of the text has left a remnant of a much more perfect code. If that were true, the surviving ELSes should almost all have short skips.

The authors also remark that "the total amount of divergence from the original text has probably been enough to obliterate any perfect pattern several times over, not merely to dilute it."

At least Drosnin presents an argument, flawed though it is. Edgar Cayce, a forerunner of the New Age movement, doesn't bother; he just asserts. His biography is an absurdly uncritical account of his involvement with astrology, bogus cures, and a host of other pseudosciences. There's no obvious figure comparable to Cayce today, but the internet is brimming with mini-Cayces and their legions of followers.

I received emails in response to most of my columns, but this one elicited some quite vituperative ones concerning not so much Cayce as my dismissive attitude toward astrology and the channeling of responses from dead relatives. It's comforting for some, I guess, to attribute personality types to zodiac signs and to have Delphic Zoom meetings with deceased loved ones. Alas, whatever gets you through the night.

Edgar Cayce: An American Prophet—Yeah, Right

James van Praagh, John Edward, and Sylvia Browne are only the most well known of the large current crop of on-air psychics and mediums. They deliver their flapdoodle on TV with seeming sincerity and often claim to speak with the dead.

One of the dead they may now more easily commune with is their spiritual ancestor, Edgar Cayce, the subject of a huge new biography, *Edgar Cayce: An American Prophet*, by Sidney K. Kirkpatrick.

Cayce is considered by many to be the forerunner of the New Age movement for his alleged medical clairvoyance, scientific insights, and much else. If one 100th of the claims implicit in his biography were warranted, this book review would not be appearing here but rather would be trumpeted on all the network news shows and emblazoned on the front page of every newspaper in the country. Still, he was an interesting character.

Born on a Kentucky farm in 1877, Edgar Cayce was very religious, sensitive, and given to frolicking with imaginary playmates and angels. Thought to be rather peculiar even at a young age, Cayce suffered a number of strange childhood mishaps—a nail penetrating his head, a baseball thrown into his spine, and a stick piercing his testicle. Despite these unusual misfortunes, the outline of his early life is simple. He grows up, becomes a photographer, marries his hometown sweetheart, moves from one small southern city to another, starts a family, and struggles financially. Gradually, however, he becomes convinced of his mystical gifts and medical intuitions.

The author was given unlimited access to Cayce's files, and the results are unfortunate. Perhaps to generate credibility, the book relentlessly recites detail after superficial detail: apartments lived in, houses bought and sold, jobs taken, businesses invested in, financial arrangements and partners, city streets and scenes.

There are descriptions of acquaintances of all sorts—including quite tenuous connections to Edison, Woodrow Wilson, Tesla, Lindbergh, Houdini, Hemingway, Earhart—and, most of all, readings of medical cases.

The readings were analyses of people who went to Cayce (or whose stories were told to him) for medical advice. He would famously drop into a trance with the help of various facilitators and while in this state would channel whatever the "Source" said about the person's medical condition, usually concluding with a prescription for therapy, often unconventional.

Many of the readings sound very much like the nebulous prescriptions of present-day mediums. The book's completely uncritical reporting is disappointing and most exasperating. Kirkpatrick seems to reject nothing, never demurs at anything, establishes no critical distance, and provides little feel for what made Cayce tick. The good news is that eventually this approach becomes amusing, and the reader eagerly anticipates the next outlandish achievement and its straight rendering.

Kirkpatrick's idea of proof is to cite scads of testimonials, including many from doctors and celebrities. Testimonials, of course, are notoriously unreliable, and there are no discussions of statistics or methodological issues.

Apparently, no statistics on the percentage of cases cured exist, and the reader must decide whether the "cures" recorded were due to Cayce's miraculous psychic insight or to a combination of the placebo effect, natural recoveries, patient selection, good common sense, dumb luck, cold reading techniques, and vague changes counted as successes.

Not surprisingly, excuses for the failure of readings abound in the book. Indeed, Cayce couldn't save his own son or various other members of his family.

As he grew older, Cayce did not limit himself to medical readings. He consulted the Source extensively on behalf of credulous business partners interested in Texas oil wells, the stock market, horse races, and even Hollywood screenplays. All of his get-rich-quick schemes failed, and he retreated once again to medical readings and less falsifiable prophecies.

Still, he never met a pseudoscience he didn't like and was an ardent believer in astrology, reincarnation, perpetual motion machines, the fabled city of Atlantis, and prophetic dreams. Moreover, his beliefs, visions, and readings were bizarrely interconnected. The reason, for example, for the technological advances of the present age is that many people living today are reincarnations of the technologically savvy denizens of Atlantis.

Kirkpatrick tells us that Cayce had the astonishing ability to lay his head on a book and thereby absorb its contents without formally reading it. As I slogged through this ungainly, preposterous, and absurdly detailed book, I found myself longing for the same facility.

The book does have one use, however. You can throw it at your TV when psychics start relaying silly messages from viewers' dead relatives.

As noted, there has been a resurgence of religiosity in recent years, especially in politics. It's not easy to be a candidate for political office without being obviously, conventionally, and enthusiastically religious. Hence a suggestion for politicians: When you occasionally invoke the inclusive nature of American society and go through the litany of welcoming Christians, Jews, and Muslims to some event, go a natural step further and welcome people of all religious persuasions as well as nonbelievers. The number of atheists and agnostics in this country is hard to measure, especially since most of these increasingly many millions of Americans don't advertise, but a politician's greater inclusiveness might even pay political dividends. It's also the right thing to do.

Unfortunately, however, things seem to be moving in a more sectarian direction. It's not difficult to imagine laws like the one discussed below being passed here and elsewhere and, what's worse, actually being enforced. In fact, such laws already have been. At times, it seems that the Enlightenment values of reason, individualism, and skepticism are being revoked.

Happily, Ireland's blasphemy law was repealed in 2020.

MEDIEVAL BLASPHEMY LAW JUST PASSED IN MODERN IRELAND
When a modern Western country whose economy is based on science and technology adopts an absurdly medieval law, one would think that this would be a news story of at least moderate size.

Oddly, though, almost no attention has been paid in the United States to the passing last month of a bill establishing a crime of blasphemy in Ireland. Approved by the Irish parliament, it states, "A person who publishes or utters blasphemous matter shall be guilty of an offence and shall be liable upon conviction on indictment to a fine not exceeding 25,000 euro."

Furthermore, "a person publishes or utters blasphemous matter if (a) he or she publishes or utters matter that is grossly abusive or insulting in relation to matters held sacred by any religion, thereby causing outrage among a substantial number of the adherents of that religion, and (b) he or she intends, by the publication or utterance of the matter concerned, to cause such outrage."

Even if I weren't the author of a book titled *Irreligion: A Mathematician Explains Why the Arguments for God Just Don't Add Up*, I would find this bill abysmally wrongheaded.

Although it provides for exceptions to prosecution if a "reasonable person" finds literary, scientific or other significant value in a work, it would allow for atheists to be prosecuted for denying the existence of God, a denial that clearly causes outrage in many.

Those writing parodies and bad jokes would also be liable to the 25,000-euro fine. Even an innocuous riff on God rescinding the Bible in the middle of the night the way Amazon called back the Orwell book from its Kindle reader could be prosecuted.

And if the reaction of some irate readers of my aforementioned book is any indication, so could an imagined instant message exchange between me and God that appears in the book.

But nonbelievers would not be the only or even the primary ones affected by this blasphemy bill. People, irreligious or not, presumably could be prosecuted for drawing cartoons of Muhammad. Christians could be prosecuted for expressing scorn or even disbelief in the Christian teachings of other denominations.

Likewise, Jews and others could be prosecuted for denying the divinity or even the existence of Jesus. Or, if atheism is considered a religion (which it is not), atheists also could claim to be outraged by the expressions of their religious countrymen, each of whom could then be required to cough up 25,000 euros.

The law also allows for the confiscation of blasphemous materials—novels, nonfiction books, short videos, full-length movies, and so on.

Interestingly, the blasphemy law is not the only medieval aspect of Irish law. The preamble to the Irish constitution maintains that the state's authority derives from the most holy trinity, stipulates that no one can become president or a judge without taking a religious oath, and declares that all citizens have obligations to Our Lord Jesus Christ.

Similar but less overt sentiments and statutes exist in this country. Witness the arguments put forth by many that the United States is a Christian country.

More analogous is a little-known example involving the state of Arkansas, which has not yet roused itself to rescind article 19 of its constitution: "No person who denies the being of a God shall hold any office in the civil departments of this State, nor be competent to testify as a witness in any court." A few other states have similar laws.

The impulse to enact benighted laws of this sort gives rise to more than these Taliban-like religious laws. After all, it is not only all-mighty deities that need special legal protection. Generals and politicians do too, so the same fearful defensiveness also leads to draconian edicts to protect political leaders and parties from ridicule.

Pakistan, to cite a recent example, has just announced a prohibition of jokes about President Asif Zardari. Anyone sending emails, text messages, or blog postings containing such jokes is subject to arrest and a 14-year prison sentence. I'm sure even more prohibitive restrictions exist in those hotbeds of freewheeling political humor, Burma and North Korea.

It's instructive to contrast these authoritarian laws against blasphemy, jokes, political humor, and free speech generally with the way people deal with dissent from established scientific laws. No laws prohibit people from denying that Earth is spherical, that evolution explains the development and diversity of life, or that the moon landing ever took place. The same holds for mathematics. No one claiming that pi is a rational number, that there are finitely many prime numbers, or that Godel's theorem is false has ever been hauled into court.

Of course, I don't mean to equate the irreligious with scientific quacks. Just the opposite in fact. It's simply that in most domains, those who insist on denying conventionally accepted beliefs are for the most part simply ignored. Statements that can stand on their own two feet (evidence and logic) don't need crutches (blasphemy laws) to support them.

As mentioned, Ireland is a modern pluralistic state with an educated population, a world-class literary tradition, and a healthy economy that has transformed itself in recent years in large part through science and high-tech jobs. To continue this transformation, the religious and irreligious alike should reject this silly blasphemy law.

The religious should probably be most opposed to it, however. Placing punitive sanctions on the robust or even the rude expression of irreligious thought does not seem to say much for religion.

That there simply cannot not be some order or law in any situation is an insight that I've always found to be profound. Put differently, a complete lack of order is impossible. One example: The random movement of molecules in a gas gives rise, on a higher level of analysis, to the laws of thermodynamics.

RAMSEY THEORY AND THE PROFOUND IDEA OF ORDER FOR FREE

Despite those who believe there is a reason, whether religious or otherwise, behind every bit of complexity in the world, it is not at all unusual for order to arise naturally. I think this rather profound realization should be more widely appreciated.

Since writing my book *Irreligion* and some of my recent Who's Counting columns, I've received a large number of emails from subscribers to creation science (who have recently christened themselves intelligent design theorists). Some of the notes have been polite, some vituperative, but almost all question "how order and complexity can arise out of nothing."

Since they can imagine no way for this to happen, they conclude there must be an intelligent designer, a God. (They leave aside the prior question of how He arose.)

My canned answer to them about biological order talks a bit about evolution, but they often dismiss that source of order for religious reasons or because of a misunderstanding of the second law of thermodynamics.

Because the seemingly inexplicable arising of order seems to be so critical to so many, however, I've decided to list here a few other sources for naturally occurring order in physics, math, and biology. Of course, order, complexity, entropy, randomness, and related notions are clearly and utterly impossible to describe and disentangle in a column like this, but the examples below from *Irreligion* hint at some of the abstract ideas relevant to the arising of what has been called "order for free."

Necessarily Some Order

Let me begin by noting that even about the seemingly completely disordered, we can always say something. No universe could be completely random at all levels of analysis.

In physics, this idea is illustrated by the kinetic theory of gasses. There, an assumption of disorder on one formal level of analysis, the random movement of gas molecules, leads to a kind of order on a higher level, the relations among variables such as temperature, pressure, and volume known as the gas laws. The law-like relations follow from the lower-level randomness and a few other minimal assumptions. (This bit of physics does not mean that life has evolved simply by chance, a common mischaracterization of evolution but with a smidgeon of truth.)

There is also the Coase theorem in law and economics, which can be interpreted as order arising naturally in many common contexts.

In addition to the various laws of large numbers studied in statistics, a notion that manifests a different aspect of this idea is statistician Persi Diaconis's remark that if you look at a big enough population long enough, then "almost any damn thing will happen."

Ramsey Order

A more profound version of this line of thought can be traced back to British mathematician Frank Ramsey, who proved a strange theorem. It stated that if you have a sufficiently large set of geometric points and every pair of them is connected by either a red line or a green line (but not by both), then no matter how you color the lines, there will always be a large subset of the original set with a special property. Either every pair of the subset's members will be connected by a red line or every pair of the subset's members will be connected by a green line.

If, for example, you want to be certain of having at least three points all connected by red lines or at least three points all connected by green lines, you will need at least six points. (The answer is not as obvious as it may seem, but the proof isn't difficult.)

For you to be certain that you will have four points, every pair of which is connected by a red line, or four points, every pair of which is connected by a green line, you will need 18 points, and for you to be certain that there will necessarily be five points with this property, you will need between 43 and 55 points. With enough points, you will inevitably find uni-colored islands of order as big as you want, no matter how you color the lines.

A whole mathematical subdiscipline has grown up devoted to proving theorems of this same general form: How big does a set have to be so that there will always be some subset of a given size that will constitute an island of order of some sort?

Ramsey-type theorems are probably also part of the explanation (along, of course, with Diaconis's dictum) for some of the equidistant letter sequences that constitute the Bible codes. Any sufficiently long sequence of symbols, especially one written in the restricted vocabulary of ancient Hebrew, is going to contain subsequences that appear meaningful.

Self-Organization and Order

Finally, of more direct relevance to evolution and the origin of living complexity is the work of Stuart Kauffman. In his book *At Home in the*

172

Universe, he discusses what he has termed the aforementioned notion of "order for free."

Motivated by the idea of hundreds of genes in a genome turning on and off other genes and the order and pattern that nevertheless exist, Kauffman asks us to consider a large collection of 10,000 lightbulbs, each bulb having inputs from two other bulbs in the collection.

Assume that you connect these bulbs at random, that a clock ticks off one-second intervals, and that at each tick, each bulb either goes on or off according to some arbitrarily selected rule. For some set of 10,000 bulbs, the rule might be to go off at any instant unless both inputs are on the previous instant. Another rule might be to go on at any instant if at least one of the inputs is off the previous instant. Given the random connections and random assignment of rules, it would be natural to expect the collection of bulbs to flicker chaotically with no apparent pattern.

What happens, however, is that very soon, one observes order for free, more or less stable cycles of light configurations, different ones for different initial conditions. Kauffman proposes that some phenomenon of this sort supplements or accentuates the effects of natural selection.

Although there is certainly no need for yet another argument against the seemingly ineradicable silliness of "creation science," these lightbulb experiments and the unexpected order that occurs so naturally in them do seem to provide one.

It has also become increasingly clear that very simple relationships or laws (such as those discussed in physicist Stephen Wolfram's *New Science* or those defining the intricately filigreed shape known as the Mandelbrot set) can give rise to complexity upon complexity.

In any case, order for free and apparent complexity greater than we might naively expect are no basis for believing in God as traditionally defined. Of course, we can always redefine God to be an inevitable island of order or some sort of emergent mathematical entity. If we do that, the above considerations can be taken as indicating that such a pattern will necessarily exist, but that's hardly what people mean by God.

This column sketches the gist of a most clever proof of Godel's heralded incompleteness theorem first developed by Greg Chaitin. Like Godel's theorem, it shows the limitation inherent in any formal system of axioms, but it hinges on a mathematical approach to complexity rather than relying on self-reference alone.

CHAITIN ON COMPLEXITY, RANDOMNESS, AND INEVITABLE INCOMPLETENESS

Some things are simple, some are complicated. What makes them so? In fields ranging from biology to physics to computer science, we're often faced with the question of how to measure complexity. There are many possible measures, but one of the most important is algorithmic complexity. First described by two mathematicians, the Russian Andrei Kolmogorov and the American Gregory Chaitin, it has been extensively developed by Chaitin in recent years.

The flavor of the subject can perhaps be sampled by considering this question: *Why is it that the first sequence of 0s and 1s below is termed orderly or patterned and the second sequence random or patternless?* (Note that since almost everything from DNA to symphonies to this very column can be encoded into 0s and 1s, this is not as specialized a question as it may at first appear.)

(A) 001001001001001001001001001001001001010 . . .

(B) 1000101101101100010101100101111010010111010 . . .

Answering this question leads not only to the definition of algorithmic complexity but also to a better understanding of (a type of) randomness as well as a proof of the famous incompleteness theorem first proved by the Austrian mathematician Kurt Godel.

Hang on. The ride's going to be bumpy, but the view will be bracing.

With sequences like those above in mind, Chaitin defined the *complexity of a sequence of 0s and 1s to be the length of the shortest computer program that will generate the sequence.*

Let's assume that both sequences continue on and have lengths of 1 billion bits (0s and 1s). A program that generates sequence A will be essentially the following succinct recipe: print two 0s, then a 1, and repeat this x times. If we write the program itself in the language of 0s and 1s, it will be quite short compared to the length of the sequence it generates. Thus, sequence A, despite its billion-bit length, has a complexity of, let's say, only 10,000 bits.

A program that generates sequence B will be essentially the following copying recipe: first print 1, then 0, then 0, then 0, then 1, then 0, then 1, then . . . ; there is no way any program can compress the sequence. If we write the program itself in the language of 0s and 1s, it will be at least as long as the sequence it generates. Thus, sequence B has a complexity of approximately 1 billion bits.

We define a sequence to be *random if and only if its complexity is (roughly) equal to its length*, that is, if the shortest program capable of generating it has (roughly) the same length as the sequence itself. Thus, sequence A is not random, but sequence B is.

Chaitin has employed the notions of complexity and randomness to demonstrate a variety of deep mathematical results. Some involve his astonishing number omega, which establishes that chance lies at the very heart of deductive mathematics. These results are explained in many of his books, including the forthcoming *Meta Math*. Let me just briefly sketch his proof of Godel's theorem.

He begins with a discussion of the Berry sentence, first published in 1908 by Bertrand Russell. This paradoxical sentence asks us to consider the following task: *"Find the smallest whole number that requires, in order to be specified, more words than there are in this sentence."*

Examples such as "number of hairs on my head," "the number of different states of a Rubik's cube," "the number of orderings of 10 decks of cards," and "the speed of light in millimeters per decade" each specify, using no more than the 20 words in the given sentence, some particular whole number. The paradoxical nature of the task becomes clear when we realize that the Berry sentence itself specifies a particular whole number that, by its very definition, the sentence contains too few words to specify. (Not quite identical but suggestive is this directive: You're a short person on an elevator in a very tall building with a bank of buttons before you, and you must press the first floor that you can't reach.)

What yields a paradox about numbers can be modified to yield mathematical statements about sequences that can be neither proved nor disproved. Consider a formal mathematical system of axioms, rules of inference, and so on. Like almost everything else, these axioms and rules can be systematically translated into a sequence of 0s and 1s, and if we do so, we get a computer program P.

We can then conceive of a computer running this program and over time generating from it the theorems of the mathematical system (also encoded into 0s and 1s). Stated a little differently, the program P generates sequences of 0s and 1s that we interpret as the translations of mathematical statements, statements that the formal system has proved.

Now we ask whether the system is complete. Is it always the case that for a statement S, the system either proves S or proves its negation, not S?

To see that the answer to this question is no, Chaitin adapts the Berry sentence to deal with sequences and their complexity. Specifically, he shows that the following task is also impossible (though not paradox inducing) for our program P: *"Generate a sequence of bits that can be proved to be of complexity greater than the number of bits in this program."*

P cannot generate such a sequence since any sequence that it generates must, by definition of complexity, be of complexity less than P itself is. Stated alternatively, there is a limit to the complexity of the sequences of 0s and 1s (formal translations of mathematical statements) generated (proved) by P. That limit is the complexity of P, considered as a sequence of 0s and 1s, and so statements of complexity greater than P's can be neither proved nor disproved by P. You simply can't prove 10-pound theorems with 5 pounds of axioms.

This is a thumbnail of a thumbnail of Godel's incompleteness theorem. It can be interpreted to be a consequence of the limited complexity of any formal mathematical system, a limitation affecting human brains as well as computers.

The above leaves out almost all the details and is, I realize, almost impenetrably dense if you have never seen this kind of argument. It may, however, give you a taste of Chaitin's profound work.

One last question: What if there were a theory of everything that was beyond our complexity horizon, as would, of course, be likely. Why do we think we'd understand a theory of everything or even be able to achieve a full understanding of our own brains? After all, a program P can only generate sequences of complexity less than that of P itself. A quote from computer scientist Emerson M. Pugh is relevant: "If the human brain were so simple that we could understand it, we would be so simple that we couldn't."

Finally, a brief excerpt from my book *Irreligion*, published by Basic Books, a whimsical counterweight to the instances of religious dogmatism mentioned above.

MY DREAMY INSTANT MESSAGE EXCHANGE WITH GOD(DESS)

I'm an atheist. If you torture me, I might say I'm an agnostic, but the torture would have to be really painful and quite extended. Nevertheless, I dreamt I had a cryptic instant message exchange with a rather reasonable and self-effacing entity who claimed to be God. This is my reconstruction of our conversation.

> *Me:* Wow, you say you're God. Hope you don't take offense if I tell you that I don't believe in you?
>
> *God:* No, that's fine. I doubt if I'd believe in me either if I were you. Sometimes I even doubt if I believe in me, and I am me. Your skepticism is bracing. I'm afraid I don't have much patience for all those abject believers who prostrate themselves before me.
>
> *Me:* Well, we share that sentiment, but I don't get it. In what sense are you God, aside from your email address—god@universe.net? Are you all-powerful? all-knowing? Did you have something to do with the creation of the universe?
>
> *God:* No, No, and No, but from rather lowly beginnings I have grown more powerful, I've come to understand more, I've emerged into whatever it is I am, and I know enough not to pay much attention to nonsensical questions about the "creation" of the universe.
>
> *Me:* It's interesting that you claim to be God yet use quotation marks to indicate your distance from the writings of some of those who believe in you.

God: I already told you that I'm a little tired of those people. I didn't create the universe but gradually grew out of it or, if you like, evolved from the universe's "biological-social-cultural" nature. How about that hyphenated word for quotation marks? You might guess that the quotation marks suggest that sometimes I want to distance myself from some of my own writings.

Me: I like that you're no literalist. Any evidence of irony or humor seems to me to be a good sign. Still I'm not sure I understand. Are you saying that you sort of evolved out something simpler, maybe something like us humans?

God: I guess you could say that except for the fact that my background is much more inclusive than that of just you humans. And looking around at what a mess you've made, I'm tempted to say, "Thank God for that," but that seems a little too self-congratulatory. Besides you've done a lot of good things too, and I've had my share of failure and misadventure, and I'm still learning.

Me: So, you're a bit of an underachieving comedian? And I take it you're a natural being, not a supernatural one?

God: Well, yes and no. I'm natural in the sense that any explanation of my provenance, existence, and slow development would be a scientific one. I'm supernatural only in the sense that I'm really rather super. That's not to say I'm super because I aspire to—excuse the term—lord it over you. It's just a straightforward statement of fact that along many dimensions (but not all) I've come to a greater understanding of things than you yet have. So it might be more accurate to say I'm relatively super.

Me: Relatively super but still a relative. A bit mightier but not almighty. Right?

God: Those are nice ways of putting it.

Me: And believers? As relatively super, you probably see them as pretty ignorant, maybe something like the cargo cultists of the Pacific, picking things to worship without any sort of natural context or much real understanding.

God: No, I'm more kindly disposed towards them than that. In fact, I love the poor benighted "souls." That last word is intended figuratively, of course.

Me: I'm still confused. Are you, despite being a bit mightier, ever confused about things? Are you ever torn in different directions, not completely certain?

God: Oh my God, yes. I'm regularly confused, torn, and uncertain about all manner of things. I can't measure up to all that perfect God stuff. Makes me feel inferior. Whatever was that Anselm thinking? For example, I wish I could constrain the most superficially ardent of my believers and tell them to cool it. Look around and think a bit. Marvel at what you've come to understand and endeavor to extend your scientific understanding. Then again I think, no, they have to figure this out for themselves.

Me: If you're as knowledgeable as you claim, why don't you at least explain to us lower orders the cure for cancer, say.

God: I can't do that right now.

Me: Why not? You can't intervene in the world?

God: Well, the world is very complicated, so I can't do so yet in any consistently effective way. Still, since I'm actually a part of the world, any future "interventions," as you call them, would be no more mysterious than the interventions of a wise anthropologist on the people he studies, people who in turn might influence the anthropologist. Nothing miraculous about entities affecting each other. Nothing easy about predicting the outcomes of these interactions either, which is why I'm hesitant about interfering.

Me: You've declared you're advanced in many ways, but do you claim to be unique? Do other "Gods" or other "a bit mightiers" a bit mightier than you exist? Do other "super universes" exist? See, I can use quotation marks too. And where are you? In space? inherent in other sentient beings? part of some sort of world-brain?

God: Not sure what questions like this even mean. How do you distinguish beings or universes? And in what sense do you mean "exist"? Exist like rocks, like numbers, like order and patterns, or maybe like the evanescent bloom of a flower? As I said, I'm not even sure I'm God, nor would I swear that you aren't. Maybe God is our ideals, our hopes, our projections, or maybe you humans are all super simulations on some super yearn engine like God-gle.

Me: The Matrix, the dominatrix, the whatever. Hackneyed, no? Anyway, even if you do exist in some sense, and I'm not buying that, you're certainly nothing like God as conventionally conceived. Do you think there is a God of that sort?

God: I know of no good evidence or logical argument for one.

Me: I agree there, but I also suspect most people would find you a pretty poor substitute for that God.

God: That's tough, just too, too bad. Something like me is the best they're going to get, and that's if they get anything at all. But as I said, I'm not positive about any of this, so let's forget the God blather for now. If I had a head, I'd have a headache. What do you say?

Me: Okay, Thy will be done, if you say so. Let's just listen to some music, assuming you have ears on your nonexistent head.

God: Yeah. (God laughs.)

Me: Yeah. (I wake up.)

CHAPTER SEVEN

Mathematically Flavored Books

I have always imagined that Paradise will be a kind of library.
—JORGES LUIS BORGES

THROUGHOUT THE YEARS, I'VE WRITTEN A NUMBER OF BOOK REVIEWS
for a variety of publications ranging from the *New York Times* and the
Washington Post to the *American Scholar* and the *New York Review of
Books*. The few I've included below are ones I wrote for my ABCNews.
com column, Who's Counting. The books chosen are not books on mathematics but do, I think, give a hint of the richness of the kind of thinking
mathematics inspires.

In fact, what is often lacking in discussions of mathematical and
other technical notions is the metamathematics necessary to make sense
of them. By "metamathematics," I don't mean the formal study of mathematical structures and properties by means of mathematics itself, which
is the usual definition. The question whether a particular mathematical
system is consistent or complete, for example, is a metamathematical
question and can be investigated mathematically.

The term "metamathematics," however, can also have a looser, more
informal meaning that refers to the context, limitations, applications, and
general place of the mathematics in question in the lay of the cognitive
land. These latter concerns in general are not mathematical ones but ones
that demand stories and extended narratives to clarify. On this understanding of the term, a metamathematical account of a theorem or other

bit of mathematics would address the concerns above by setting the full scene for it, even by providing a sort of biographical sketch of it.

Some of the books discussed below do something like this, as do books about mathematics and related areas by a variety of engaging contemporary authors. Supplementing mathematical texts with such metamathematical (in the latter sense) discussions and digressions would probably change the view of many that mathematics is simply a collection of arcane techniques of one sort or another.

"Sticks and stones may break my bones, but names will never hurt me." It turns out that sticks and stones may have more to do with the origin of mathematical concepts than more formal theories.

METAPHORS-R-US AND THE ORIGIN OF MATHEMATICAL CONCEPTS

I hear it all the time: Mathematics is impossibly esoteric. You're born with mathematical talent, or you're not. One solves math problems instantaneously. The source of mathematical insight is unfathomable. And so on and on.

The movie *A Beautiful Mind* tells the fascinating story of mathematician John Nash, but unfortunately it also suggests to many that the aforementioned beliefs are true.

It may not be the intent of the recently released *Where Mathematics Comes From* to combat these widespread misconceptions, but happily that is one of its effects.

The book's authors, linguist George Lakoff and psychologist Rafael Nunez, analyze the cognitive basis of mathematical ideas and in the process suggest new avenues of educational research.

So where does mathematics come from? Not surprisingly, none of us start out with a knowledge of differential equations. Instead, the authors contend that from a rather puny set of inborn skills—an ability to distinguish objects, to recognize very small numbers at a glance, and, in effect, to add and subtract numbers up to 3—people extend their mathematical powers via an ever-growing collection of metaphors.

Our common experiences of standing up straight, pushing and pulling objects, and moving about in the world lead us to form more complicated ideas and to internalize the associations among them.

In fact, the authors argue that we understand most abstract concepts by projecting our physical responses onto them. The notion of a conceptual metaphor is well known from Lakoff's earlier work, particularly *The Metaphors We Live By*, a book that underscored how metaphors pervade our everyday thinking about the world. Physical warmth, for example, helps elucidate our understanding of affection: "She was cool to him." "He shot her an icy stare." "They had the hots for each other."

Lakoff and Nunez take a metaphor to be an association between a familiar realm, something like temperature, construction, or movement, and a less familiar one, something like arithmetic, geometry, or calculus. The size of a collection (of stones, grapes, toys), for example, is associated with the size of a number. Putting collections together is associated with adding numbers and so on.

Another metaphor associates the familiar realm of measuring sticks (small branches, say, or pieces of string) with the more abstract one of arithmetic. The length of a stick is associated with the size of a number once some specified segment is associated with the number 1. Scores and scores of such metaphors underlying other more advanced mathematical disciplines are then developed.

Demystifying Mathematical Ideas

Throughout the book, the authors attempt to demystify mathematical thought. (Recall Eugene Wigner's statements about the unreasonable effectiveness of mathematics.) They stress that mathematical ideas do not gush out of some pipeline to the Truth (such as John Nash's schizophrenia) but have a source similar to that of other, more prosaic notions. The root of some of our mystification, they argue, is the "numbers equals things" metaphor, which leads to the Platonic idea that numbers are "up there" somewhere. Lakoff and Nunez are intent on debunking this belief and others linked to it.

The second half of *Where Mathematics Comes From* is a bit more technical and deals largely with infinity and the metaphors that animate our understanding of the ideas in calculus, such as limits and infinite series. Ultimately, Lakoff and Nunez return to the nature of the existence of mathematical objects. Whether they are mental constructions, facets of an idealized reality, or just rule-governed manipulations (like the game of chess) is an issue that has resonated all through the history of philosophy and is certainly not settled in this book, although it's an approach I find quite compelling.

Whatever one's views on the nature of mathematical entities and truths, however, the book is provocative and beneficial in its emphasis on the metaphorical aspects of mathematical concepts. A deeper appreciation of the sometimes unconscious, usually mundane sources of mathematical ideas can only help us learn and teach mathematics.

To demonstrate this, the book ends with an extended case study of the authors' approach to mathematical idea analysis. In it, they clearly explain all the layers and interconnections among the metaphors necessary to develop an intuitive grasp of Euler's famous equation, $e^{\pi i} + 1 = 0$, relating five of the most significant numbers in mathematics.

Bertrand Russell wrote of the "cold, austere beauty" of mathematics. In very different ways, *A Beautiful Mind* and *Where Mathematics Comes From* remind us of the warm bodies from which this beauty arises

Although many might view the linking of mathematics and narrative the way they view the linking of fish and bicycles, the connection is multifaceted. One of my books, *Once Upon a Number*, is an attempt, among many others, to bridge the seeming chasm between numbers and narratives or, to vary the alliteration, between stories and statistics.

MATHEMATICS AND NARRATIVE: A MULTIFACETED RELATIONSHIP
At first glance (and maybe the second one too), narrative and mathematics don't seem to be natural companions, but recent years have made the juxtaposition much more common. There have, for example, been many biographies about mathematicians ranging from Sylvia Nasar's

A Beautiful Mind about John Nash to Rebecca Goldstein's just released *Incompleteness: The Proof and Paradox of Kurt Godel*.

There have been many narrative accounts of mathematical ideas and theorems as well, ranging from Simon Singh's *Fermat's Enigma: The Epic Quest to Solve the World's Greatest Mathematical Problem* to a spate of tomes on the Riemann hypothesis by Karl Sabbagh, John Derbyshire, Marcus Du Sautoy, and others.

There have also been dramatic renditions of mathematical ideas or mathematicians in works such as David Auburn's *Proof*, Tom Stoppard's *Arcadia*, John Barrow's *Infinities*, and Apostolos Doxiadis's *Uncle Petros and the Goldbach Conjecture*. There's even a new television murder mystery show, *Numb3rs*, featuring a crime-solving mathematician. (This latter reminds me of a joke that generally appeals only to mathematicians: How do you spell Henry? Answer. Hen3ry. The 3 is silent.) And these just scratch the surface. Countless—well, not really, you can count them— narrative renderings of things mathematical have poured forth in recent years.

Arguably, even books such as *The Da Vinci Code* that are about neither mathematics nor mathematicians derive some of their appeal from mathematical elements within them. So, I think, does much humor, but that's another story.

From the Narrative of Math to the Math of Narrative
With all this ferment, it's perhaps not surprising that the phenomenon has attracted academic interest. Scheduled for July 12–15 in Mykonos, Greece, an international conference on mathematics and narrative will explore the interplay between these two seemingly disparate ways of viewing the world. There will be mathematicians (among them, myself), computer scientists, writers, and none-of-the-aboves who will examine the math–narrative nexus from many different perspectives.

In addition to looking at the narrative of mathematics, some contributors will discuss the mathematics of narrative, utilizing mathematical ideas to study narrative techniques. Hypertext, recursion theory, combinatorics, and other mathematical notions can no doubt shed light on these techniques.

The underlying logic of narrative and mathematics is different in many ways, and this will also be explored. In mathematics, for example, equals can always be substituted for equals without changing the truth-value of statements. That is, whether we use 3 or the square root of 9 in mathematical contexts doesn't affect the truth of our theorem or the validity of our calculation. Not so in narrative contexts. Lois Lane knows that Superman can fly, but she doesn't know that Clark Kent can fly even though Superman equals Clark Kent. The substitution of one for the other can't be made. Similarly, former president Reagan may have believed, as the apocryphal story has it, that Copenhagen is in Norway, but even though Copenhagen equals the capital of Denmark, it can't be concluded by substitution that Reagan believed that the capital of Denmark is in Norway.

The symbiotic way in which mathematical metaphors elucidate everyday speech and, conversely, the way in which mathematical notions derive from everyday stories and activities is of interest as well. Notions of probability and statistics, for example, did not suddenly appear in the full dress regalia we encounter in mathematics classes. There were glimmerings of these concepts in stories dating from antiquity, when bones and rocks were already in use as dice. References to likelihood appear in classical literature, and the importance of chance in everyday life must have been understood by at least a few. It's not hard to imagine thoughts of probability flitting through our ancestors' minds. "If I'm lucky, I'll get back before they finish eating the beast."

Many other mathematical topics might be mentioned, but suffice it to say that an increasing appetite for abstraction as well as other aspects of contemporary culture will ensure that the confluence of mathematics and narrative will intensify.

I'll end with this plaint from Lewis Carroll, who unfortunately will not be in attendance:

> Yet what mean all such gaieties to me
> Whose life is full of indices and surds
> $X^2 + 7X + 53 = 11/3$

The Mandelbrot set is one of those dizzying fractal shapes that have come to symbolize the modern science of chaos theory. Its characteristic filigreed intricate pattern seems to be everywhere.

FRACTAL FIND: MONK DISCOVERED MATHEMATICAL FORMULA 700 YEARS EARLIER THAN PREVIOUSLY THOUGHT

An astonishing instance of mathematical anticipation has recently come to light, showing that a monk discovered a complex formula centuries earlier than previously thought.

The Mandelbrot set is one of those dizzying fractal shapes that have come to symbolize the modern science of chaos theory. Like all such shapes, it is indefinitely convoluted and has arcs giving rise to smaller arcs and flares branching into smaller flares.

Oriented vertically, it can be viewed as a star, and it is this way of seeing the set that leads to what is probably one of the most arresting discoveries in the history of mathematics.

On his wonderfully multitudinous website, The Apothecary's Drawer, Roy Girvan, an English science writer and Web designer, tells the story of a retired mathematics professor traveling in Germany who was shocked to see in a medieval religious manuscript a nativity scene with a Star of Bethlehem in the unmistakable shape of the Mandelbrot set.

The professor, Bob Schipke, confessed to being stunned. "It was like finding a picture of Bill Gates in the Dead Sea Scrolls. The title page named the copyist as Do of Aachen, and I just had to find out more about this guy."

After investigation, Schipke discovered that this 13th-century monk, previously known only for his poetry and essays, had discovered the simple secret for the generation of the Mandelbrot set 700 years before Benoit Mandelbrot!

Happily, some of Do's notebooks have survived, and Schipke found that the genius monk wrote knowledgeably about the notion of probability. He also discovered that the monk had discovered a method for estimating the value of pi that had previously been thought to have been first employed by 18th-century naturalist Comte de Buffon.

But Do of Aachen's most staggering feat is the construction of the Mandelbrot set, which involves the multiplication of complex numbers. These are numbers of the form a + bi, where a and b are what we normally think of as real numbers and i, a so-called imaginary number, is the square root of −1.

Spiritual, Profane, and Complex

Incredibly, these numbers would not become a part of mathematical practice for another 500 years, but Do of Aachen wrote of the intertwined spiritual and profane parts of all entities, including numbers, and through some scrim of theological reasoning, he concluded that the product of two purely profane numbers was a negative spiritual number, just as $i^2 = -1$.

In this way, Do derived rules for dealing with profane/spiritual numbers that were essentially identical to those used today for manipulating complex numbers.

Using these rules and more theology (a number multiplied by itself is somehow the analogue of spiritual reflection) and repeatedly substituting complex numbers into the expression $z^2 + c$ to determine their "fate" eventually led Do of Aachen to the Mandelbrot set.

Professor Schipke, with the help of historian Antje Eberhardt at the University of Munich, wrote up his findings on Do in the March 1999 issue of the *Harvard Journal of Historical Mathematics*.

But despite the prestige of the publication, the discovery has not received the attention it warrants, and this may be one reason that Ray Girvan has so beautifully told the story on his website and is certainly why I am publicizing it here.

Not since the Piltdown–Sokal theorem was discovered to have first been proved in 1907 by a provincial Russian electrical engineer and not in 1979 by Piltdown and Sokal as originally had been thought has the date of a first proof been so misjudged.

There is only one case of misattribution in the history of mathematics that is of comparable magnitude. The famous triangular array of numbers discovered by Pascal in the 17th century was anticipated five centuries before him by the Chinese mathematician Chia Hsien.

Even this example, however, is trumped by the 700-year differential between Do's and Mandelbrot's discovery.

But seriously . . .

P.S. It's interesting that all the complexity in the Mandelbrot set comes from such a simple equation that it could have been discovered by a 13th-century monk who stumbled upon rules for dealing with imaginary numbers.

Incidentally, the final adjective ("imaginary") applies as well to Girvan's Do of Aachen himself. Don't believe everything you read. Especially on this date, April 1.

In *Number Sense*, Stanislaus Dehaene argues that numerical sense is innate. Since his book, a number of studies have strengthened and extended his contention.

THE MATHEMATICAL BRAIN: WE ALL HAVE AN INNATE SENSE OF NUMBERS AND MAGNITUDES

The new school year looms ominously for many who claim to lack a "mathematical brain," and so it may be a good time to review the findings of Brian Butterworth and Stanislaus Dehaene, two cognitive psychologists who have done much work on the neural basis of mathematical thinking. Both their books, Butterworth's *What Counts: How Every Brain Is Hardwired for Math* and Dehaene's *The Number Sense: How the Mind Creates Mathematics*, maintain that there are in the left parietal lobe of the brain certain specialized circuits that enable us to do arithmetic.

These circuits, which Butterworth terms the Number Module, ensure that all of us can automatically recognize very small numbers, match up the objects in small collections, and tell which of two small collections is larger. We do these tasks unthinkingly the way we note colors without trying to do so. Furthermore, any numerical achievements beyond this are a result of our slowly mastering various representations of numbers supplied by the surrounding culture. These include body parts (fingers primarily), external aids such as tallies and abaci, and written symbols such as Roman or Arabic numerals.

Other cultural tools, laboriously discovered throughout the centuries and presented in classrooms this fall, enable us to master more advanced mathematical notions such as algebra, probability, and differential equations.

We certainly differ in the extent to which we master these tools, but we all start with the same basic mathematical brain, the authors argue.

To support their thesis that numerical notions are a part of our innate neural hardware, Butterworth and Dehaene describe experiments in which researchers present babies with white cards that have two black dots on them.

They place the cards a few inches from the babies' eyes and note how long the babies stare at them. The babies soon lose interest but resume staring when the researchers show them cards with three black dots.

After the babies lose interest in these cards, they regain it only when shown cards with two dots again. The babies appear to be responding to the change in number since they seem to disregard changes in the color, size, and brightness of the dots. Another experiment: When researchers place two dolls behind a screen in front of babies but only one remains when they remove the screen, the babies are surprised. The researchers elicit a similar surprise when they place one doll behind the screen and there are two when they remove the screen.

The babies are not surprised if two dolls turn into two balls or a single doll turns into a single ball. The conclusion is that violations of quantity are more disturbing to babies than changes in identity.

Disorders Can Tell Us More

Victims of disorders in the brain's number module and their resulting deficits provide more support for claims about the region. There have been many such cases.

A person has a stroke that damages the left parietal lobe and, although still articulate, can no longer tell without counting whether there are two or three dots on a sheet of paper. Someone can't say what number lies between 2 and 4 but has no problem saying what month is between February and April. Someone else with a tumor in the left parietal lobe cannot connect the arithmetic facts she recites in a singsong way to any real-world application of them. One patient understands arithmetic procedures but can't recall any arithmetic facts, while another has the opposite condition.

Particularly intriguing is Gerstmann's syndrome, which is characterized by finger agnosia (an inability to identify particular fingers upon request) and acalculia (an inability to calculate or do arithmetic). Butterworth theorizes that during a child's development, the large area of the brain controlling finger movements becomes linked to the specialized circuits of the Number Module, and the fingers come to represent numbers.

In arguing for the innateness of some numerical concepts, both authors take exception to the work of the Swiss psychologist Jean Piaget. In one of Piaget's famous experiments, for example, researchers showed very young children two identical collections and then moved the objects in one collection farther apart.

The children were likely to say that the spread-out collection had more objects, and Piaget concluded they did not yet really understand the notion of quantity. More recent experiments seem to show that what the children did not understand was the question they were being asked. Dehaene shows that if the same children are asked to choose between four jelly beans spread apart and five jelly beans close together, they are very unlikely to go for the four jelly beans.

Of course, education is still important, and since the number module is hardwired in all of us, Butterworth and Dehaene argue that one of the primary reasons (sometimes they implausibly appear to be saying the only reason) for disparities in mathematical achievement is environmental—better instruction, more exposure to mathematical tools, motivation for hard work.

The authors note the burden imposed on students by the cumulative nature of mathematical ideas and the self-perpetuating nature of different attitudes toward the subject. In particular, Butterworth contrasts the virtuous circle of encouragement, enjoyment, understanding, and good performance leading to more encouragement with the vicious circle of discouragement, anxiety, avoidance, and poor performance leading to more discouragement.

There is much else of interest in both of these books, but my sense of number tells me I've gone on for long enough.

The mathematical concept of infinity is illustrated in popular writings, including a novel and a play by the two prominent authors mentioned below as well as in many other nonliterary ways. Part of the reason, of course, is that people are naturally drawn to the notion of eternity and the certainty of at least something going on and on, even if it's only the Platonic fantasm of numbers.

INFINITY AND ETERNITY: A NOVELIST'S MATH, A PHYSICIST'S DRAMA

Numbers and narratives, statistics and stories. From Rudy Rucker's *Spaceland* to Apostolos Doxiadis's *Uncle Petros and Goldbach's Conjecture*, from plays such as *Copenhagen*, *Proof*, and *Arcadia* to many nonstandard mathematical expositions, the evidence is building. There has always been some interplay between mathematics and literature, but the border areas between them appear to be growing. Increasingly, fiction seems to come with a mathematical flavor, mathematical exposition with a narrative verve.

Two recent works on the mathematical notion of infinity illustrate this phenomenon. One is by novelist David Foster Wallace, the author of the exuberant 1,088-page novel *Infinite Jest*, among other works of fiction. His new book *Everything and More: A Compact History of Infinity*, which I reviewed for *The American Scholar*, sketches the history of humanity's attempts to understand infinity. It begins with the Greeks and ends with modern logicians, Georg Cantor in particular. In between are accounts of the attempts by many mathematicians to get a handle on the discombobulating notions of the infinitely big and the infinitesimally small.

The other work on the topic is a play titled *Infinities* by English physicist and cosmologist John Barrow. So far performed only in Europe, the play dramatically explores various counterintuitive aspects of infinity, from a scenario devoted to Jose Luis Borges's parable of the Library of Babel to one about the implications of mathematician David Hilbert's Hotel Infinity.

To get a feel for the latter, imagine a scenario in which you arrive at a hotel, hot, sweaty and impatient. Your mood is not improved when the clerk tells you that they have no record of your reservation and that the hotel is full. "There is nothing I can do, I'm afraid," he intones officiously. If you're in an argumentative frame of mind and know some set theory, you might in an equally officious tone inform the clerk that the problem is not that the hotel is full but rather that it is both full and finite.

You can explain that if the hotel were full but infinite (the afore-mentioned Hilbert's Hotel Infinity), there would be something he could do. He could tell the guest in room 1 to move into room 2; the original party in room 2 he could move into room 3, the previous occupant of room 3 he could move into room 4, and so on. In general, the hotel could move the guest in room N into room (N + 1). This action would deprive no party of a room yet would vacate room 1, into which you could now move.

Infinite hotels clearly have strange logical properties, and they don't stop there. Infinite hotels that are full can find room not only for one extra guest but also for infinitely many extra guests. Assume the infinitely many guests show up at Hotel Infinity demanding a room. The clerk explains that the hotel is full, but one of the extra guests suggests the following way to accommodate the newcomers. Move the guest in room 1 into room 2 and the guest in room 2 into room 4. Move the guest in room 3 into room 6 and, in general, the guest in room N into room 2N. All the old guests are now in even-numbered rooms, and the infinitely many new guests can be moved into the odd-numbered rooms.

In a sense, this property of infinity has been known since Galileo, who pointed out that there are just as many even numbers as there are whole numbers. Likewise, there are just as many whole numbers as there are multiples of 17. The following pairing suggests why this is true: 1–17, 2–34, 3–51, 4–68, 5–85, 6–102, and so on.

Where physicist Barrow's play relies on a sort of abstract drama to get its ideas across, novelist Wallace's book has a more conventional for-mat and covers more mathematical ground. Among many mind-bend-ing oddities, it discusses different orders of infinity; in a quite precise sense, the set of all decimal numbers is "more infinite" than the set of all fractions, which is "no more infinite" than the set of all whole numbers. Wallace also discusses Georg Cantor's unprovable continuum hypothesis, which deals with these various orders of infinity.

The task Wallace has chosen is heavy going, but he brings to it a refreshingly conversational style as well as a reasonably authoritative command of mathematics. Because the language is smart and inventive, the book provides enough enjoyment to induce the mathematically unsophisticated reader to slog through the many difficult patches along the way.

And this is part of the value of the confluence of these two works and of others like them. Although it's extremely unlikely that a novelist will prove a new theorem and only slightly less improbable that a mathematician will write a great novel, these attempts to span the chasm between the so-called two cultures should be applauded. Mathematical exposition is too important to be left only to mathematicians, and the wide variety of literary forms available should not be off limits to mathematicians and physicists.

Maybe one day Heartbreak Hotel will be just down the creative block from Hotel Infinity, but right now, before I succumb to a further spasm of cutesiness, it's time to check out.

Gould describes some surprising connections between batting averages and bacterial complexity as well as between IQ and statistics. As with most writing on mathematics broadly speaking, it remains apposite despite being at times superficially somewhat dated. So says the author of this book.

STEPHEN GOULD'S USE OF MATH CLARIFIES HIS INSIGHTS ON BASEBALL, BACTERIA, AND IQ

Stephen Jay Gould, the eminent evolutionary biologist and stylish essayist who has since died, used mathematics to elucidate many ideas, both in and out of science.

Gould's was a wide-ranging and impassioned voice of reason that will be keenly missed. One aspect of his work that particularly appealed to me was his use of simple mathematical observations and analogies to help clarify his multifarious arguments.

Gould was, for example, famously interested in baseball, bacteria, and the complexity of life and characteristically managed to connect them in an enlightening and nontrivial way. Consider his explanation for the disappearance of the .400 hitter in baseball in his book *Full House: The Spread of Excellence*.

He argued that the absence of such hitters in recent decades was due not to any decline in baseball ability but rather to a gradual decrease in the disparity between the worst and best players, both pitchers and hitters. When almost all players are as athletically gifted as they are now, the distribution of batting averages shows less variability. The result is that .400 averages are now very scarce. Players' athletic prowess is close to the "right wall" of ultimate human excellence in baseball.

Leading to bacteria and complexity, Gould next asked his readers to consider an imaginary country in which initially every adult receives an annual salary of $100. Assume that every few years, each person's salary is either adjusted upward by $100 (say, 45% of the time) or downward by the same amount (say, 55% of the time) with the proviso that no one's salary ever declines below $100, the minimum wage.

After a number of generations, the largest salary in the country will likely be quite a bit larger than $100, and the average will rise somewhat as well. This is because there is, at first, only one direction for salaries to grow; there is a "left wall" below which salaries can't decline. Although the $100 salary becomes less common throughout time, it nonetheless remains the most common salary.

And bacteria and complexity? Gould was interested in reconciling the obvious growth in the complexity of organisms throughout time with the Darwinian contention that the complexity of highly developed species is as at least as likely to decrease as it is to increase. His resolution is the following. Bacteria, the first life form on Earth, have minimal complexity, roughly analogous to the $100 minimum salary of the imaginary country above.

The apparent trend in overall complexity growth is a consequence of the fact that no life forms are simpler than bacteria, just as a rise in average salary follows from the fact that no one can make less than the $100 minimum wage in the imaginary country. Random changes in biological complexity can initially be in only one direction even though, in more developed species, decreases in complexity are at least as likely to occur as increases. Evolution provides no inherent drive for species to become more complex; global biological progress is an illusion. (Gould's argument is, of course, much more nuanced than a brief description can capture.)

Medians, Cancer, and I.Q.s

A more heartening use of a mathematical notion appeared in his well-known article "The Median Isn't the Message." There, Gould discussed being diagnosed in 1982 with abdominal mesothelioma, a cancer whose median survival time is (or at least was) eight months; that is, half the people with it live less than eight months, half longer. He took comfort from the fact that the mean or average survival time might nevertheless be considerably longer than eight months. This could occur in the same way that the average home price in a neighborhood might be $1,000,000 even if the median price were just $50,000 (if, say, there were a few palatial mansions surrounded by many quite modest homes).

Gould's book *The Mismeasure of Man* is also full of mathematical observations, most devoted to undermining the foundations of IQ testing. He criticized disguised confusions of correlation and causation in the misapplication of statistical techniques such as factor analysis and regression analysis. In a related *New Yorker* piece on the book *The Bell Curve*, he illuminated the IQ debate with a discussion on the heritability of height in impoverished villages in India. Tall fathers there, say, taller than 5 feet, 7 inches, generally have tall sons, while short fathers there, say, shorter than 5 feet, 2 inches, generally have short sons.

Despite this undisputed heritability of height (much more so than for IQ), most would conclude that improved nutrition and sanitation would raise the average height of the villagers to that in the West. The conclusion: The heritability of characteristics (such as height or IQ) within populations is not an explanation for average differences between populations.

Being an evolutionary biologist, Gould was always conscious not merely of averages but also of variations, exceptions, and diversity. Appropriately, his writings across a wide range of disciplines were themselves various, exceptional, and diverse.

He was a rare throwback to a less myopically specialized era.

The practice of nudging is interesting but troubling. For example, it could perhaps be useful in reducing virulent political partisanship by somehow linking moderate reality-based websites to counter ones that are clearly nonsensical. Doing so, however, would likely be perceived as liberal paternalism. Another possibility that has just been tried is to use people's exaggerated estimate of their chances of winning a lottery to combat their exaggerated estimate of their chances of having a serious reaction to the COVID vaccine. Get vaccinated, and you'll be eligible for a specially crafted lottery. Jiu Jitsu for the innumerate. I must admit I still have reservations about the uncritical exploitation of people's gullibility with nudges.

Nudging: How to Get People to Do the Right Thing—Maybe

Nudge, a book by cognitive psychologist Richard Thaler and legal scholar Cass Sunstein that came out last year, puts forward a simple thesis. Because people often behave unthinkingly, it's better in some cases to lure (or nudge) them into making the right choice rather than trying to convince them of its rightness and/or imposing legal sanctions against the wrong choices.

To achieve this end, common psychological foibles can be used, as can appropriate "choice architectures" and default options.

The urinals at Schiphol airport in Amsterdam provide an odd introductory example of this. In an effort to reduce the amount of spillage on the floor, small depictions of houseflies were embossed on the porcelain at the base of the urinals. These "fly targets," which have since been adopted throughout the world, reduced spillage by 80% without the necessity of written notices, larger basins, or other expensive interventions.

A similarly nondirective nudge results from taping a picture of human eyes over an unattended receptacle meant to collect money on the honor system from the sale of, let's say, doughnuts or coffee.

Yet another is the placement of food in a school cafeteria line. If healthful foods are placed first, they're more likely to be chosen than if desserts or fried foods are at the beginning of the line.

Other cognitive lapses also give rise to nudges. "Save to Win," a savings program devised by Harvard Business School professor Peter Tufano, is a more recent instance. It relies on people's exaggerated estimation of their likelihood of winning a lottery.

A response to Americans' perennially low savings rate, the program induces the members of eight credit unions in Michigan to buy certificates of deposit (CDs) by automatically enrolling purchasers in a lottery. The monthly drawings pay out $400 and the annual one $100,000. Bigger organizations would no doubt have bigger jackpots. The CD pays a slightly lower rate of interest than do standard CDs, presumably to pay for the lottery.

A more common example of exploiting people's innate propensities is the practice of activists who focus on individuals with whom the public can identify rather than on numbers that can give a better sense of the scope of the problem. If the goal is to nudge people into donating money or time, much recent research indicates that the individual-based approach leads to larger contributions and greater commitment to action than does mention of the numbers.

Thus, if the issue is the trafficking of people across national borders, it's more effective to focus on a single young woman and her incarceration in a foreign brothel or on a single young man and his entrapment on a foreign farm or construction site than it is to cite a bunch of numbers. In fact, including the dreaded and impersonal numbers often lessens interest.

Investment decisions might also benefit from what Thaler and Sunstein call "libertarian paternalism." People confronted with too many choices, say, for their 401(k) plans, often choose unwisely. They don't diversify, don't consider the tax consequences of their holdings, don't consider low-cost index funds, and so on.

If companies offered just as many choices but went through the trouble of carefully devising a very good single default plan that requires no action, most people would succumb to inertia and go with it. This seems to have been the result in Sweden when that country revamped its retirement plans.

Consideration of the appropriate-choice architecture might very well help with the drafting of health care legislation. Structuring the immensely complex system is an extraordinarily difficult job that requires not only a free market (the mother of all nudges) but also a wise choice of extra-market incentives. In their book, Thaler and Sunstein consider a number of related problems.

Should the pricing of insurance policies be structured so that people who partially give up their right to sue because of negligence pay less for health care? What about inducements to lose weight, go to a gym, and so on? Are there nudges that might increase organ donations?

Despite its appeal, providing nudges and sometimes even tricking people into doing the right thing, even when it's clear what "right" is, makes me very uncomfortable.

I've come to accept the necessity of this, however, because such tricks have been used for so long by more than a few politicians, advertisers, credit card companies, health insurers, lobbyists, and others to tempt people into doing what is, by almost any reasonable standard, the wrong thing.

In fact, special interests often go beyond mere nudges into distracting spin, tendentious framing of issues, and even bald-faced lies. It's telling, for example, that frightening talk about trillion-dollar health care reform and phony death panels rarely seemed to arise about the trillion-dollar Iraq War and the real deaths to which it did lead.

It's a sad fact that rhetoric convinces far more people than does logic. Given that it does, intelligent nudging, as long as it doesn't degenerate into stupid fudging, is to be cautiously welcomed. At least arguably.

Physicist Stephen Wolfram's book *A New Kind of Science* looks at science as a collection of recursive rules that he claims are capable of generating all the complexity we see in the world. Wolfram suggests that his rule

110 should take its place alongside Turing machines, general recursive functions, Conway's *Game of Life*, and a number of such idealized universal computers.

WOLFRAM'S *NEW KIND OF SCIENCE*: SIMPLE RULES CAN GENERATE ALL THE COMPLEXITY WE SEE

People tend to engage in a form of primitive thinking known as "like causes like," psychologists have shown. For example, doctors once believed that the lungs of a fox cured asthma and other lung ailments. People assumed that fowl droppings eliminated the similar-appearing ringworm. Freudians asserted that fixation at the oral stage led to preoccupation with smoking, eating, and kissing.

It is perhaps not surprising, therefore, that people have long thought that the complexity of computer outputs was a result of complex programs. It's been known for a while, however, that this is not necessarily the case. Computer scientists and mathematicians, notably John von Neumann in the 1950s and John Conway in the 1970s, have studied simple rules and algorithms and have observed that their consequences sometimes appear inordinately complex.

Nevertheless, it's safe to say that no one has treated this idea with anything like the intensity and thoroughness of Stephen Wolfram in his fascinating, ambitious, and controversial new book, *A New Kind of Science*.

Simple Rule, Complex Consequences

The book is a mammoth 1,200 pages, so let me in this 800-word column focus on Wolfram's rule 110, one of a number of very simple algorithms capable of generating an amazing degree of intricacy and, in theory at least, of computing anything any state-of-the-art computer can compute.

Imagine a grid (or, if you like, a colossal checkerboard), the top row of which has some white squares and some black ones. The coloring of the squares in the first row determines the coloring of the squares in the second row as follows:

Pick a square in the second row and check the colors of the three squares above it in the first row (the one behind it to the left, the one

200

immediately behind it, and the one behind it to the right). If the colors of these three squares are, respectively, WWB, WBW, WBB, BWB, or BBW, then color the square in the second row black. Otherwise, color it white. Do this for every square in the second row.

Via the same rule, the coloring of the squares in the second row determines the coloring of the squares in the third row, and, in general, the coloring of the squares in any row determines the coloring of the squares in the next row. That's it, and yet, Wolfram argues convincingly, the patterns of black and white squares that result are astonishingly similar to patterns that arise in biology, chemistry, physics, psychology, economics, and a host of other sciences. These patterns do not look random, nor do they appear to be regular or repetitive. They are, however, (in some senses) exceedingly complex.

Moreover, if the first row is considered the input and black squares are considered to be 1s and white ones 0s, then each succeeding row can be considered the output of a computation that transforms one binary number into another.

Not only can this simple so-called one-dimensional cellular automaton perform the particular calculation just described, but, as Wolfram proves, it is capable of performing all possible calculations! It is a "universal" computer that, via appropriate coding, can emulate the actions of any other special-purpose computer, including, for example, the word processor on which I am now writing.

A number of such idealized universal computers have been studied (ranging from Turing machines through John Conway's *Game of Life* to more recent examples), but Wolfram's rule 110 is especially simple. He concludes from it and myriad other considerations too numerous and detailed to even adumbrate here that scientists should direct their energies toward simple programs rather than equations since programs are better at capturing the complicated interactions that characterize scientific phenomena.

A New Kind of Science also puts forward a "Principle of Computational Equivalence," which asserts, among other things, that almost all processes, artificial (such as his rule 110) or natural (such as those occurring in biology or physics), that are not obviously simple can give rise to universal computers. This is reminiscent of an old result known as the Church–Turing thesis, which maintains that any rule-governed process or computation that can be performed at all can be performed by a Turing machine or an equivalent universal computer. Wolfram, however, extends the principle, gives it a novel twist, and applies it everywhere.

In fact, his polymathic approach is part of what is exciting about the book. Simple programs, he avers, can be used to explain space and time, mathematics, free will, and perception as well as help clarify biology, physics, and other sciences. They also explain how a universe as complex appearing and various as ours might have come about: The underlying physical theories provide a set of simple rules for "updating" the state of the universe and such rules are, as Wolfram demonstrates repeatedly, capable of generating the complexity around (and in) us if allowed to unfold during long enough periods of time.

Some of the book's claims are hubristic and hyperbolic. Many others have been around for a while, but no one has put them all together in such a compelling way to articulate if not a new kind of science at least a new way of looking at the established kind.

A Book with a Theory of Almost Everything? The book includes uncompromising discussions of space-time and Minkowskian geometry, general relativity theory, Lagrangian and Hamiltonian approaches to dynamics, quantum particles, entanglement, the measurement problem, Hermitian operators, black holes, the Big Bang, time travel, quantum field theory, the anthropic principle, Calabi–Yau spaces, as well as many other topics of current research.

It may be a book with a theory of almost everything, but it's most decidedly not a book for almost everybody.

PENROSE'S *ROAD TO REALITY*: A BEHEMOTH ON MODERN MATH AND PHYSICS—READER CAUTION ADVISED

There's an oft-repeated story that when Stephen Hawking was writing *A Brief History of Time*, he was told that every equation in the book would cut his readership in half.

If there were any truth to this counsel, Roger Penrose's *The Road to Reality: A Complete Guide to the Laws of the Universe*, his recent 1,100-page behemoth of a book, should attract a half dozen readers at most. It's an enormous equation-packed excursion through modern mathematics and physics that attempts, quixotically perhaps, to answer and really explain "What Laws Govern Our Universe?"

Scattered about this impressive book are informal expository sections, but Penrose's focus is on the facts and theories of modern physics and the mathematical techniques needed to arrive at them. He doesn't skimp on the details, which, for different readers, is the book's strength and its weakness. Parts of it, in fact, seem closer in tone to a text in mathematical physics than to a book on popular science.

An emeritus professor at Oxford, Penrose is a mathematician and physicist renowned for his work in many areas.

In the 1960s, he and Hawking did seminal research on "singularities" and black holes in general relativity theory. He also discovered what have come to be called Penrose tiles, a pair of four-sided polygons that can cover the plane in a nonperiodic way. And about a decade ago, he wrote *The Emperor's New Mind*, in which he argued that "artificial intelligence" was a bit of a crock and that significant scientific advances would be needed before we could begin to understand consciousness.

Mathematical Preliminaries: Fractions to Fiber Bundles

The first 400 pages of *The Road to Reality* sketch the mathematics needed to understand the physics of the following 700 pages. Like many mathematicians, Penrose is an avowed Platonist who believes that mathematical entities such as pi, infinite cardinal numbers, and the Mandelbrot set are simply "out there" and have an objective existence independent of us.

Developing his mathematical philosophy a bit with some interesting speculations about the relations between the mathematical, physical, and mental worlds (but never descending to sappy theology), he very soon gets into the mathematical nitty-gritty. He expounds on Dedekind cuts, conformal mappings, Riemann surfaces, Fourier transforms, Grassmann products, tensors, Lie algebras, symmetry groups, covariant derivatives, and fiber bundles, among many other notions.

As suggested, the level of exposition and the topics covered make me wonder about the intended audience. Penrose writes that he'd like the book to be accessible to those who struggled with fractions in school, but this seems an almost psychotically optimistic hope. This is especially so because his approach to so many topics is so clever and novel.

Another problem is that he doesn't generally proceed from the concrete to the theoretical but more often in the other direction. (For the mathematicians: He introduces abstract 1-forms and only later the relatively more intuitive vector fields. Likewise, he develops Maxwell equations via tensors and Hodge duals and never explicitly mentions more familiar notions like the curl of a field or Stokes' theorem.) The physics begins around page 400 and includes uncompromising discussions of space-time and Minkowskian geometry, general relativity theory, Lagrangian and Hamiltonian approaches to dynamics, quantum particles and entanglement including the standard illustrations (the two-slit experiment, Schrodinger's cat, and the Einstein–Podolosky–Rosen nonlocality), the measurement problem, Hermitian operators, black holes, the Big Bang, time travel, quantum field theory, the anthropic principle, Calabi-Yau spaces, as well as many other topics of current research.

EPR Experiment, String Theory, Inflation, and Everything Else

As in his previous works, the author is not afraid to strike an iconoclastic pose. He sides with Einstein and against most modern physicists, for example, in thinking that the EPR experiment demonstrates that quantum theory is incomplete.

The experiment, described very simplistically since this column has fewer words than Penrose's book has pages, involves identical particles moving rapidly apart. A physicist measures the spin of one of the particles, realizing that quantum theory stipulates that the particle doesn't have a definite spin—it could go either way—until it is measured and its wave function collapses. Astonishingly, the other particle, which by the time of the measurement may be in a different galaxy, has a wave collapse at the same moment that always results in its having an opposite spin. How does the second particle instantaneously "know" the first particle's spin? Eerie entanglement, an incomplete theory, something else?

Penrose's skepticism extends to more modern developments as well.

He is unenthused about inflation theory and particularly so about string theory. (Inflation, very roughly, refers to the lightning-fast expansion of a part of the very early universe, and string theory, even more roughly, refers to the notion that fundamental particles are composed of minuscule strings, vibrating and multidimensional.) Inflation theory has considerable evidence backing it, but Penrose seems correct to emphasize that string theory and its offspring M-theory are largely speculative. Why their appeal? He offers an interesting discussion of the role of fads and fashion even in theoretical physics.

The end of the book is devoted to a sketch of M-theory's main competitor, twistor theory and loop quantum gravity, which he invented decades ago and has been developing with colleagues ever since. He also seems less than impressed with Brian Greene's *The Elegant Universe*, a book that is far more accessible.

Coming every few pages, Penrose's well-done drawings and illustrations may ease the book's near vertical learning curve. Like some *New Yorker* subscribers, many readers of this book will, I suspect, confine themselves largely to the pictures and the pages that are more broad gauged and less technical.

There is something to be said for inducing even this level of involvement in mathematics and physics, and if *The Road to Reality* succeeds in doing this, it may become a (sturdy) coffee-table book and popular success. My hunch, however, is that this truly magisterial book will be appreciated primarily by those who have already spent considerable time in school learning a substantial portion of what's in it.

Is catastrophe in our cards? A mathematical way to assess the risk of doomsday looks at our reign on Earth as a lottery. This is perhaps a fitting way to end this book. Or perhaps not.

IS THE SKY FALLING? AND IF SO, WHEN? A PROBABILISTIC DOOMSDAY ARGUMENT

Constant reports about pandemic diseases, environmental catastrophes, societal collapse, and nuclear proliferation revive these perennial human questions about possible (or likely) existential threats and contribute to our feelings of anxiety and unease.

So too did the recent passing of an asteroid almost 100 feet in diameter within 30,000 miles of the Earth. Such news stories make a recent abstract philosophical argument a bit more real.

Developed by a number of people, including Oxford philosopher Nick Bostrom and Princeton physicist J. Richard Gott, the *Doomsday Argument* (at least one version of it) goes roughly like this.

There is a large lottery machine in front of you, and you're told that in it are consecutively numbered balls, either 10 of them or 10,000 of them. The machine is opaque, so you can't tell how many balls are in it, but you're fairly certain that there are a lot of them. In fact, you initially estimate the probability of there being 10,000 balls in the machine to be about 95%, of there being only 10 balls in it to be about 5%.

Now the machine rolls, you open a little door on its side, and a randomly selected ball rolls out. You see that it is ball number 8, and you place it back into the lottery machine. Do you still think there is only a 5% chance that there are 10 balls in the machine?

Given how low a number 8 is, it seems reasonable to think that the chances of there being only 10 balls in the machine are much higher than your original estimate of 5%. Given the assumptions of the problem, in fact, we can use a bit of mathematics called Bayes' theorem to conclude that your estimate of the probability of 10 balls being in the machine should be revised upward from 5% to 98%. Likewise, your estimate of the probability of 10,000 balls being in it should be revised downward from 95% to 2%.

What does this have to do with doomsday? To see, let's imagine a cosmic lottery machine, which contains the names and birth orders of all human beings from the past, the present, and the future in it. Let's say we know that this machine contains either 100 billion names or, the optimistic scenario, 100 trillion names.

And how do we pick a human at random from the set of all humans? We simply consider ourselves; we argue that there's nothing special about us or about our time and that any one of us might be thought of as a randomly selected human from the set of all humans, past, present, and future. (This part of the argument can be much more fully developed.)

If we assume there have been about 80 billion humans so far (the number is simply for ease of illustration), the first alternative of 100 billion humans corresponds to a relatively imminent end to humankind—only 20 billion more of us to come before extinction. The second alternative of 100 trillion humans corresponds to a long, long future before us.

Even if we initially believe that we have a long, long future before us, when we randomly select a person's name from the machine and the person's birth order is only 80 billion or so, we should re-examine our beliefs. We should drastically reduce or, so the argument counsels, our estimate of the likelihood of our long survival, of there ultimately being 100 trillion of us.

The reason is the same as in the example with the lottery balls: The relatively low number of 8 (or 80 billion) suggests that there aren't many balls (human names) in the machine.

Here's another slightly different example. Let's assume that Al receives about 20 emails per day, whereas Bob averages about 2,000 per day. Someone picks one of their accounts, chooses an email at random from it, and notes that the email is the 14th one received in the account that day. From whose account is the e-mail more likely to have come?

There are many other examples devised to shore up the numerous weak points in the doomsday argument. Surprisingly, many of them can be remedied, but a few of them, in my opinion, cannot be.

That a prehistoric man (who happened to understand Bayes' theorem in probability) could make the same argument about a relatively immi-nent extinction is an objection that can be nicely addressed. Appealing to some so-called anthropic principle whereby inferences are drawn from the mere fact that there are observers to draw them is much more prob-lematic. In any case, there's probably still time to learn more about the doomsday argument and the use of the so-called anthropic principle in philosophy, cosmology and even everyday life. A good place to begin is Nick Bostrom's work, particularly his book *Anthropic Bias*.

The end?

Postscript

If people do not believe that mathematics is simple, it is only because they do not realize how complicated life is.

—John von Neumann

Messy complications necessarily infect uses of mathematics in everyday life. On the other hand, they also provide the chaotically bubbling energy of life. We should be skeptical of them, but we should also prize them for what they might teach us.

The writings herein illustrate, I hope, some of the problems and the promises of such uses and display the perennial tension between the abstract and the topical. How to strike a balance between timeless statistics and timely stories is the last puzzle in this book, and I leave it to the reader.

My take on the balance, at least in matters of societal significance, is to focus more on the quantitative and the abstract rather than on fleeting details and empty fluff. On the other hand, I hate the facile "mathematical applications" that often result from a mindless focus on numbers.

However we strike the balance, we should be mindful, as I've stressed before, that mathematics is no more computation than literature is typing. It deserves a rightful place in our never-ending endeavor to make some sense of the world and, whenever possible, to try to better it.

Q.E.D.